Pesticide Selectivity

Pesticide Selectivity

edited by Joseph C. Street

Department of Animal Science
Utah State University
College of Agriculture
Logan, Utah

MARCEL DEKKER, INC. New York and Basel

MARCEL DEKKER, INC.

270 Madison Avenue, New York, New York 10016

LIBRARY OF CONGRESS CATALOG CARD NUMBER: 75-12365

ISBN: 0-8247-6335-1

Current printing (last digit):
10 9 8 7 6 5 4 3 2 1

PRINTED IN THE UNITED STATES OF AMERICA

CONTENTS

Contributors v

Preface vii

1. Broad-Spectrum and Narrow-Spectrum Herbicides -- A Need for Both 1

   Glenn C. Klingman

2. Biological, Chemical, and Other Factors Influencing Strategies for Selective Weed Control 11

   Robert P. Upchurch

3. Biological and Biochemical Considerations in the Use of Selective and Nonselective Fungicides 21

   Malcolm R. Siegel

4. Practical Considerations in Use of Selective Insecticides Against Major Crop Pests 47

   T. F. Watson

5. Strategies in the Design of Selective Insect Toxicants 67

   Robert M. Hollingworth

6. Target-Specific Pesticides: An Industrial Case History 113

   Gustave K. Kohn

7. Comparative Biotransformation as a Forecaster of the Ecological Consequences of Selective Insecticides 135

Tsutomu Nakatsugawa and Robert R. Stewart

8. Basis for Selectivity of Acaricides 155

Charles O. Knowles

Author Index 177

Subject Index 185

CONTRIBUTORS

Hollingworth, Robert M., Department of Entomology, Purdue University, West Lafayette, Indiana

Klingman, Glenn C., Plant Science Field Research, Eli Lilly and Company, Greenfield, Indiana

Knowles, Charles O., Department of Entomology, University of Missouri, Columbia, Missouri

Kohn, Gustave K.,* Chevron Chemical Company, Ortho Division, Richmond, California

Nakatsugawa, Tsutomu, Department of Entomology, State University College of Environmental Science and Forestry, Syracuse, New York

Siegel, Malcolm R., Department of Plant Pathology, University of Kentucky, Lexington, Kentucky

Stewart, Robert R., Department of Entomology, State University College of Environmental Science and Forestry, Syracuse, New York

Upchurch, Robert P.,** Agricultural Research Department, Monsanto Company, St. Louis, Missouri

Watson, T. F., Department of Entomology, University of Arizona, Tucson, Arizona

---

*Presently affiliated with the United Nations Industrial Development Organization, New Delhi, India

**Presently Chairman, Department of Plant Sciences, University of Arizona, Tuscon, Arizona

PREFACE

During recent years the intense public controversy over use of pesticides, revolving largely around preoccupation with the quality of environment, has frequently generated the suggestion that more narrowly selective pesticides should be developed. This view carries the implicit concept that the closer one approaches in use a one pest-one chemical pesticide balance, the greater the degree of safety to all other organisms. Minimum persistency is frequently coupled to this recipe for an environmentally acceptable pesticide, since the compound which self-destructs in the shortest period of time and is target-selective should be even more environmentally desirable.

The concept of insecticide selectivity was introduced to the scene by W. E. Ripper in 1944 as a "chemical that kills the uneconomic arthropod species and spares the economic species, namely, the pest's natural enemies."[1] Nicotine offered the earliest known example of such selectivity. Progress in developing such chemicals was not outstanding, however, so that by 1956 Ripper, in a classic review, could indicate only a few synthetic chemicals meeting his definition for selectivity.[2] Schradan was one found effective at that time. Ripper emphasized that

---

[1] W E. Ripper, Nature 153:448, 1944.

[2] W. E. Ripper, Ann. Rev. Entomology 1:403, 1956.

ecosystem stability depended on selective and partial control of the pest population by chemicals allowing the natural enemy population to stabilize on the reduced food supply represented by the pest numbers maintained near the economic threshold.

The developing physiological and biochemical understanding of insecticide toxicology soon provided an outline of the essential principles for selective insecticide action in relation to chemical properties of the compounds.[3] By the early sixties, therefore, it appeared that a rational search for development of selective insecticides should be possible. The desirability of such a search was then clearly in the public mind because of the forceful disclosures of problems with some broad-spectrum persistent insecticides as presented by Rachel Carson and other popular writers.

A clear statement of this philosophy became part of the U.S. federal research policy following the report in 1963 of the President's Science Advisory Committee on "Use of Pesticides."[4] That PSAC report stated, among others, the following recommendation: "In order to develop safer, more specific control of pests, it is recommended that government-sponsored programs continue to shift their emphasis from research on broad-spectrum chemicals to provide more support for research on (a) selectively toxic chemicals, (b) nonpersistent chemicals, (c) selective methods of application, and (d) nonchemical control methods such as attractants and the prevention of reproduction." The PSAC committee felt that production of safer, more specific, and less persistent pesticide chemicals did not represent an unreasonable goal and in this way encouraged the USDA and other to shift research programs toward development of increasingly specific controls, including selective chemicals.

---

[3] R. D. O'Brien, Adv. Pest Control Res. 4:75, 1961.

[4] President's Science Advisory Committee, <u>Use of Pesticides</u>, The White House, 1963.

After that date the rate of progress toward such goals was far from rapid with the result that the Mrak commission report to the Secretary of Health, Education, and Welfare in late 1969 recommended that "incentives should be provided to industry to encourage the development of safer chemicals with high target specificity, minimal environmental persistence, and few, if any, side effects on nontarget species."[5] By that time it was more clearly appreciated that the developmental costs of specific chemicals to be used selectively would be disproportionately high in relation to profits from the correspondingly low volume of sales for selective use. The Mrak committee thus perceived that high development costs would discourage research and development of selective pesticides without some form of incentive being provided.

Along the way it had been pointed out by E. F. Knipling, as well as others, that the development of selective systems for controlling pests cannot be accomplished without great effort and research.[6] "One of the chief advantages of a broad-spectrum pesticide is that a good one may lead to practical ways of controlling hundreds of pest species. In contrast, research on highly selective ways to control specific pests necessitates intensive research on every major pest. In many instances, the use of selective pest control measures will also mean higher cost to the grower or to the public."

The foregoing capsule history has been presented in terms of selective insecticides. Yet the various commissions and many other advocates made no distinctions and called for nonpersisting

---

[5] U. S. Dept. of Health, Education, and Welfare. Report of the Secretary's Commission on Pesticides and Their Relationship to Environmental Health.

[6] E. F. Knipling, in Pest Control Strategies for the Future, National Academy of Sciences, Washington, D. C., 1972.

selective pesticides in general, thereby including herbicides, fungicides, and all others.

This book developed from a symposium on Pesticide Selectivity held by the Division of Pesticide Chemistry, American Chemical Society, in Chicago, August 29, 1973. In presenting the symposium it was our modest intent to examine that generalization with respect to major classes of pesticides and attempt to establish a reasonable perspective. Is pest control with selective compounds uniformly feasible? Is the practical desirability of using narrowly selective herbicides similar to that for selective insecticides? What economic consequences might occur within agriculture and to the consumer through reliance on selective pesticides? Would we be following the best track in going for narrowly selective pesticides? Winteringham[7] raised this question in his recent review of insecticide selectivity in asking whether biodegradability, per se, is not a more significant quest? If economic compromising is necessary, should not this be the basis? These and related questions have been considered, if not firmly answered, in this book. The strategies found useful in obtaining selective action within classes of pesticides and progress toward achieving successful compounds are also discussed.

Joseph C. Street
Utah State University

[7]F. P. W. Winteringham, Ann. Rev. Entomology 14:409, 1969.

Pesticide
Selectivity

Chapter 1

BROAD-SPECTRUM AND NARROW-SPECTRUM HERBICIDES--A NEED FOR BOTH

Glenn C. Klingman
Plant Science Field Research
Eli Lilly and Company
Greenfield, Indiana

The topic for this chapter has as its base Recommendation 11 from the Report of the Secretary of Health, Education, and Welfare Commission on Pesticides and their Relationship to Environmental Health, perhaps better known as the Mrak report {1969}. This recommendation is quoted in part as follows:

> Provide incentives to industry to encourage the development of more specific pest control chemicals. Incentives should be provided to industry to encourage the development of safer chemicals with high target specificity, minimal environmental persistence, and few, if any, side effects on nontarget species. Developmental costs will be disproportionately high in relation to profits from the lower volume of sales of more specific chemicals which will be used selectively. The high cost of development will discourage investments unless incentives are provided.

While not a part of this chapter, let us consider for a moment the first recommendation that "Incentives should be provided to industry...." Industry does not need as "incentives" government subsidies or government assistance to do the needed research so long as it is not hampered by excessive regulatory restrictions and economic controls, and so long as it is allowed to compete in a relatively free market place. Industry does need as incentives an "appropriate climate" for the development of pesticides. This appropriate climate includes no weakening of patent procedures, and reasonable registration requirements involving meaningful research in the areas of toxicology and biological chemistry. Hazards to man and his environment must be related to real hazards, which must be researchable and provable. Through research, the exposure levels of the pesticide can be measured, and the hazards assessed. Can the applicator make the treatment without being exposed, or exposing others, or without effect to the environment, and is the food product free of harmful residues?

There is an area of government support that is needed--the research efforts and unbiased appraisal of pest control techniques by the USDA, University Research Programs, and Cooperative Extension Programs. Together with industry these programs have developed a highly productive, efficient agriculture. Food costs have been low in the past, partially at least, as a direct reflection of the success of this program.

A second recommendation concerns "the development of safer chemicals." This simple statement ignores the basic truth that most herbicides are safe and present little hazard in actual use. It would be foolhardy to say that we need "less safe" pesticides. It is not appropriate, however, to group all pesticides into one category. Each pesticide, indeed, each use of a pesticide, must be considered individually. Pesticides may be highly toxic, or far less toxic to man than common table salt.

Also, we should consider a third assumption in the Mrak recommendations--that we need "minimal environmental persistence." While possibly true with insecticides, it is not true of herbicides. Weed seeds, of one kind or another, germinate throughout a crop-growing season. If a crop is to be protected season long from weed competition, some persistence is required. For total vegetation control, such as on railroads, industrial sites, and roadsides, persistence is essential.

The important remaining part of Recommendation 11 from the Mrak report deals with favoring "narrow-spectrum pesticides" over "broad-spectrum pesticides." The remainder of this chapter, will consider the advantages and/or disadvantages of each, dealing only with herbicides.

Before proceeding, let us define the terms narrow-spectrum herbicide and broad-spectrum herbicide, taking into account that selectivity, which is mentioned by the Mrak report, is required for the use of a herbicide in a crop.

1. *Narrow-spectrum selective herbicide.* A chemical to which all plants are tolerant, except a very few. Thus, the herbicide can be applied to kill one plant or a very limited number without injury to all other plants. If such a herbicide were available, a weed such as cocklebur could be killed without injury to any crop and without effect on most other weeds. The fact of the matter is--we do not have narrow-spectrum selective herbicides on the market.
2. *Broad-spectrum selective herbicide.* A chemical that kills all plants, except a few. Thus, the herbicide can be applied to a crop (for example, cotton) and remove all other plants in the field. The chemical would be good only for cotton, but no other herbicide would be needed in cotton.

It may be of interest to note that cultivation was the first broad-spectrum method of weed control. Selective broad-sprectrum weed control was originally developed when plants were _first_ planted in rows so that a horse could drag a heavy hoe between the rows.

Most selective herbicides fit the broad-spectrum definition. Table 1 shows several of the leading herbicides, the principal crops, and the number of weeds claimed specifically on labels or supplemental labels. Note that all of these herbicides are broad-spectrum both with reference to crops and weeds.

The Weed Science Society of America published a composite list of weeds by both common and botanical names {1971}. This list contains 2,060 different weed species with 87 percent of them broadleaved weeds, 11 percent grasses, and 2 percent sedges.

TABLE 1

Trade Name, Common Name, Principal Crops, and Number of Weeds Species Claimed on the Herbicide Label or Supplemental Labels

Trade name	Common name	Crop uses	Number of weed species claimed
Aatrex	atrazine	Corn, sorghum, sugarcane, plus 4 misc. uses	40
Eptam	EPTC	Beans, potatoes, plus 6 other crop uses	34
Karmex	diuron	Cotton, sugarcane, plus 26 misc. uses	48
Lasso	alachlor	Soybeans, corn, peanuts, cotton (restricted area)	23
Treflan	trifluralin	Cotton, soybeans, plus 34 misc. crops	27

Thus, if it were possible to develop a narrow-spectrum herbicide for each species, 2,060 different herbicides would be needed.

When asked about the number of weeds that are found in a usual cultivated crop, our field researchers, as would be expected, answered that the number would dramatically vary from field to field. However, when pressed for a specific number, there was agreement that there would be an average of 10 to 12 different, but important, weed species in most cultivated crop fields. The species, however, will vary considerably from field to field, crop to crop, and from one geographical area to another.

About 20 years ago, crabgrass (_Digitaria_ sp.) was listed as the most serious weed in cultivated crop fields. It is interesting to note that crabgrass is one of the weeds controlled by all five of the important herbicides listed in Table 1. It is also interesting that crabgrass no longer is considered the number one weed. Farmers now have effective methods of controlling crabgrass in their crops.

Today, according to a number of surveys, the number one weed worldwide, is nutsedge or nutgrass (_Cyperus_ _rotundus_ L). Through the years, treatment of this weed has included the use of soil sterilant herbicides, 2,4-D, thiolcarbamate herbicides, heavy shade, and continuous clean cultivation for 2 years. However, the fact remains that none of these treatment "eradicates" the weed. With the above methods the farmer is able to control nutsedge so that he can grow a crop--but the weed quickly returns.

Let us consider then the practicality of developing a narrow-spectrum herbicide just for nutsedge control. Some of the first considerations would be the development costs and the size of the market, which will affect the cash flow of the company involved.

The average research cost of each new pesticide as given by M. B. Green at the Weed Science of America meetings {1973} was \$5.5 million, starting from synthesis through early production of experimental permit materials. Fig. 1 illustrates the cash flow of a successful pesticide. Note that during the first 6 years it is all outgo. With marketing starting in the sixth year, the red ink in the ledger does not disappear until 10-1/3 years have passed.

Assume that you are research director for a company doing pesticide development work. To develop a nutsedge herbicide, you will need to consider a number of factors. For example, perhaps some other company finds a chemical equally effective on nutsedge control, one that also controls 25 other important crop weeds. In addition, suppose that a number of important crops are tolerant to the new chemical.

After considering the facts, few of us would recommend proceeding with a separate and independent research program aimed only at the control of nutsedge. A more likely route for the development of a successful nutsedge herbicide would be from ongoing herbicide screening tests conducted by industry. By including nutsedge in this screen, the chances are increased for finding such a herbicide. In this case, it is not likely that nutsedge control, alone, will need to bear the full research

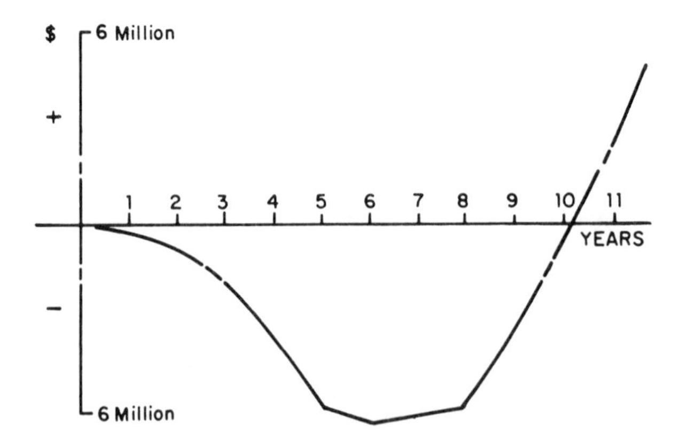

Fig. 1. Discounted cash flow of a successful pesticide.

cost of development--as it will be a broad-spectrum selective herbicide and will control other weeds in addition to nutsedge.

ENVIRONMENTAL POLLUTION EFFECTS

From a strictly environmental or pollution control point of view, let us consider narrow-spectrum vs. broad-spectrum herbicides. For discussion purposes, let us assume that there are six individual herbicides as shown in Table 2, used by soybean growers.

With the first five, one would probably need to use all five herbicides to efficiently grow soybeans. Thus, five times as much chemical could conceivably be going into the environment at a cost five times that had herbicide 6 been chosen. In addition, all the possible interactions would need to be studied, perhaps synergistic reactions as they relate to soybean tolerance or crop safety, crop residues, soil residues, efficacy of weed control, effect of the toxicology interactions on man, wildlife,

TABLE 2

The Effects, Tolerance, and Degree of Weed Control of Six Theoretical Herbicides Used in Soybean Cultivation

Herbicide	Controls	Fails to control	Tolerance	Spectrum of weed control
No. 1	Cragbrass	All other weeds	Tolerant	Narrow
No. 2	Pigweed	All other weeds	Tolerant	Narrow
No. 3	Foxtails	All other weeds	Tolerant	Narrow
No. 4	Lambsquarters	All other weeds	Tolerant	Narrow
No. 5	Johnson grass	All other weeds	Tolerant	Narrow
No. 6	All of the above plus many others	Has limited number of tolerant weeds	Tolerant	Broad

birds, fish, the environment, water contamination, etc. This interaction type of research would require a monumental effort; in fact, it would be essentially impossible. While this research was being conducted the weeds would flourish and the food supply would diminish. Thus, we cannot expect narrow-spectrum herbicides to do the full job needed in most crops.

There are limited areas where narrow-spectrum herbicides may be successful. Consider present-day herbicide programs used on corn, soybeans, or cotton using effective broad-spectrum herbicides. In addition, many of these herbicides are used in combination--to further broaden the spectrum of weed control. These herbicides are doing an excellent job on most weeds, but occasionally one or a small number of weed species escapes control. The narrow-spectrum herbicide may be successful just for the control of that specific weed.

However, the farmer, if given the choice between a narrow-spectrum and a broad-spectrum herbicide, will undoubtedly choose the broad-spectrum herbicide to serve as the backbone of his weed control program. Thus, the life of a narrow-spectrum herbicide is likely to be short.

There are monocultures of weeds in certain aquatic areas and in range areas that may justify a narrow-spectrum herbicide. However, even in such areas, the control of one such weed species would likely be followed by the immediate invasion and establishment of some other weed species.

Returning to Recommendation 11 of the Mrak report, we can summarize our evaluation as follows:

1. Incentives such as government subsidies or research assistance are not needed by industry to develop pesticides. Incentives are needed in the form of sound patent laws, and reasonable registration requirements in the areas of toxicology and biochemistry. Hazards to man and his environment must be related to real researchable hazards, not to hazards which possibly could exist.

2. Most herbicides as used are safe. Those that are not should be replaced. This should be on an individual use basis, after meaningful research.
3. Persistence may be needed to do the job. In itself, persistence is not necessarily bad.
4. If we were to shift to herbicides with high target specificity, we would dramatically increase the number of herbicides used, the total amount of herbicides used, and costs. These costs would ultimately raise the cost of food production.
5. Broad-spectrum selective herbicides are necessary because they serve as the backbone of weed control programs in agriculture.
6. Narrow-spectrum selective herbicides may serve a useful purpose wherever a single weed species is in need of control. The length of life of such a product can be expected to be short, and the size of the market will usually be small.

REFERENCES

Green, M. B., 1973. Are Herbicides Too Expensive?, Weeds Today. *4*(3):14-16.

U. S. Department of Health, Education, and Welfare, December 1969. Report of the Secretary's Commission on Pesticides and their Relationship to Environmental Health, Parts I and II, p. 17.

The Weed Science Society of America, 1971. Composite list of herbaceous weeds, aquatic weeds, and woody plants arranged alphabetically by scientific name, Weed Science. *19*(4): 437-476.

Chapter 2

BIOLOGICAL, CHEMICAL, AND OTHER FACTORS INFLUENCING STRATEGIES FOR SELECTIVE WEED CONTROL

Robert P. Upchurch
Monsanto Company
St. Louis, Missouri

Today we possess an exceptional capacity to develop strategies for selective weed control based on our experience in the past and especially in the last 25 years. The last quarter of a century has brought us the technique of weed control by herbicides. This technique has been effectively meshed with various cultural and preventive weed control measures. Prior to the introduction of the modern farm tractor and the use of herbicides, farmers struggled against weeds with highly inadequate tools. The result was that huge amounts of animal and manpower were required to combat weeds. These efforts were often unsatisfactory in that the weeds took a sharp toll in manpower, costs, and reduced productivity. In too many cases the farmer lost the battle altogether. Today the farmer battles weeds on terms which are more in his favor. The progress we have made is one of the fundamental elements in our present ability to provide ourselves with products grown on the land. This progress has been achieved with an exceptional record of providing safety to humans and their environment. As we attempt to identify strategies for the future it is only logical that we build upon what we have learned through the development and wide-scale use of modern weed control technology.

In our overall national approach to the matter of developing strategies for pest control we have fallen into a trap. The trap has been set, possibly unintentionally, by those who have concerned themselves with certain aspects of insect control and the usage of certain insecticides. From the literature and public pronouncements there seem to be four main legitimate strategies which are involved. They are:

1. Because some insecticides have as a side effect some impact on some predatory and otherwise beneficial insects, efforts should be made to minimize these side effects by developing more selective insecticides and/or by using alternate or combination techniques.
2. Because some insecticides have as a side effect some impacts of an undesirable nature on some elements of the environment, the insecticides which can be demonstrated to have such side-effects should be reduced in usage, replaced by more suitable insecticides or nonchemical measures, or used in a way that will not generate the side effects.
3. Because certain insects show a characteristic of developing resistance to certain insecticides when used at frequent intervals and at high dosages, an attempt should be made to avoid heavy reliance on this manner of using the insecticides which have this specific liability.
4. Because some insecticides influence some beneficial insects, because some insecticides have some definable, undesirable side effects in the environment, and because some insects develop resistance to some insecticides used in certain ways, the technique of integrated insect control should be developed whereby all elements of insect control can be used

on an optional basis in an optimum manner with minimum side effects and without leading to ineffectiveness in any one element. Biological control is stipulated to be a mainstay of this strategy.

As far as one can determine these four strategies are sound. Unfortunately they are most often stated in such a way that the audience gains the impression that the strategies, respectively, are as follows:

1. For various reasons broad-spectrum pesticides have severe limitations and the development of highly selective pesticides is essential if pesticides are to continue to play a role in human affairs.
2. Pesticides harm the environment and every effort should be made to reduce the usage of these materials and to eliminate their use altogether whenever and wherever possible.
3. Pest species develop resistance to pesticides and the usage of such materials has the inevitable effect of leading to the development of generations of resistant strains which are successively more difficult to control.
4. Because of the limitations and disadvantages inherent in pesticides their use should be eliminated or kept at the absolute minimum by adopting the technique of integrated pest control, with heavy reliance upon biological control.

The strategies as stated in the last four items above are illegitimate and unsound. This is because, among other reasons, they use the term "pest," which embraces all classes of pests, such as plant pathogens, weeds, nematodes, birds, etc., in a narrow sense. Unfortunately, many entomologists as well as many nonentomologists have fallen into the habit of using the words "pest" and "pesticide" when they more probably mean "insect" and "insecticide." It must be admitted that the entomologists may be concerned with pests other than insects but the usurpation of the terms pest and pesticide to apply to their unique concern is unacceptable.

It is unwise to have a national strategy on pest management unfold without due consideration of the facts relating to weeds and their control. Weed science as a profession is young and it is understaffed. Weed scientists have been devoting their attention to the more scientific and publicly less glamorous aspects of improving weed control technology. For the most part they have been out of the public eye. They have played only a minor role in the generation of national strategies on pest management and understandably may be somewhat dismayed to find themselves embraced by strategies designed by others, either intentionally or unintentionally. The time has come for weed scientists to step forward in a bolder fashion and to provide the background for the development of a national strategy on weed management. The comments herein are offered as a further contribution to whatever modest beginnings have been made in this regard.

For pest management strategy considerations it is necessary to define some fundamental differences between the classes of pests. Some contrasts between weeds and insects will suffice to provide examples. Although some weeds are readily spread, there is ordinarily a marked difference between the mobility of weed propagules and insects. Many weed species do not now exist in specific geographical localities where they could easily become a significant liability. Many species can be excluded from or restricted to specified geographical areas. On the basis of ecological principles many troublesome weed species continue to move towards the boundaries of their adaptation. Although the mobility of weeds is a burden, it is one which can be managed and our national strategy should involve containment and gradual restriction to smaller geographical areas as feasible on a species-by-species basis.

Damage done by insects frequently may be attributed to the activity of one insect species at a particular point in time and often major consideration must be given to the preservation of predator species which may hold the key to the amount of damage expected. For this reason entomologists are desirous of having and using an insecticide which is highly selective in toxic action towards a culprit insect but which is nontoxic towards predator organisms and beneficial or nonharmful guest organisms in the treated area. As an alternative the entomologists must use available but relatively nonselective insecticides in a way that will produce the desired end result.

The practitioner of weed control in crop production is normally faced with an entirely different situation than the entomologist. A field or a geographical area may contain one or several weed species which are especially pernicious and which merit particular attention. However, the general situation is that a substantial variety of species will be present, most of which must be controlled if the crop is not to be sacrificed. Also, in crop production predator relationships are not normally important. Further, herbicides, as a general rule, are highly selective between the plant and the animal kingdoms; and guest organisms of the animal kingdom in treated areas are normally not harmed except by removing plants which serve as hosts. Under these circumstances the weed scientist desires to have at his command herbicides which are well tolerated by the crop and by animals but which are completely nonselective as far as weed species are concerned. Where the herbicides available fall short of his needs, the weed scientist tries to adapt available technology to make it useful. Regardless of what the national strategy specifies for entomology, it should specify that broad-spectrum herbicides are highly desired and that utility should be determined by the features of unique selectivity towards crop plants derived either from biological or contrived sources. Often the requirements can be fulfilled best by the use of combinations of products and procedures either concurrently or successively in time and/or space.

The nature of weeds is such that it is possible to generate and use herbicides to ameliorate the damage they do without damaging man or his environment. In fact, herbicides provide a mechanism whereby the environment may be managed effectively and in a nondeleterious way for the benefit of man. The record is clear with respect to utility and safety based on herbicide usage over the last 20 years. The price of continuing to develop this sort of record is eternal vigilance. Any national effort to reduce herbicide usage should be based on the merits of specific cases. At this point in time there is adequate justification for increased attention to weeds and their control in many respects but a national effort to curtail herbicide usage based on environmental considerations would seem to be unjustified, unnecessary, and counterproductive. Reduction in usage in a given case may be warranted on the basis of economics but this is a minor consideration in the overall situation.

The fact that repeated treatment of certain insect populations with high dosages of some insecticides can lead to the development of resistant insect strains is well documented. Such a fact has practical implications and should be considered in developing national strategies for insect control. The situation is quite different in the world of weeds. In the first place weed species do not normally produce many generations per season. The conventional manner of control imposed using herbicides and other means does not give exceptionally favorable treatment to a given resistant ecotype. Reducing herbicide treatments to threshhold levels would tend to favor such herbicide resistant strains. Any resistant ecotypes that develop must contend with the specimens established from the normal weed seed reservoir in the soil in succeeding years and often under a rotational cropping or herbicide use system in which the ecotype would find no advantage. Even if the resistant ecotype establishes itself in a given field, it has the burden of spreading itself in order to create a significant impact. Whatever the reasons, it is a fact that the field of weed control has not been severly hampered by the development of herbicide-resistant weed strains. Although such entities could develop in theory it seems that any such developments will come about slowly enough so that adjustments can be made in weed control programs to offset the possible difficulties.

A more common problem in the field of weed control is that any particular combination of cultural and herbicidal practices in a given field tends to allow for the natural increase in certain weed species which have been present but which have been held in check by previous control procedures. The most striking example of this has occurred where manual techniques have been replaced by chemicals. Certain herbicidal treatments have a capacity to far outperform manual means from both efficacy and economic standpoints. Yet such treatments are minimally effective on some weed species. This is a well-recognized problem but farmers and the technical community have been making great progress in dealing with this matter. While more attention and support needs to be given to resolving these sorts of problems there is no need for a panic.

In the case of herbicides there seems to be little reason to mount a national campaign to curtail the use of herbicide technology. The treatments which have been employed for herbicidal purposes have not led to deterioration of our food supply or our environment. Our governmental regulations seem to be working effectively and the current safeguards seem to have built-in margins of safety which are reasonable and perhaps even more than reasonable. We now have governmental programs which are designed to "reduce the level of insecticide usage." This objective is frequently stated as being to "reduce the level of pesticide usage." Weed scientists should and do object strenuously to the blanket inclusion of herbicides in such programs. Their efforts in pesticide work on a national scale should be directed to those specific and real problems and opportunities which do exist in fact. There is enough worthwhile work to do without expending their resources on imaginary problems.

Concerns in the field of entomology have led to a great deal of interest in the area of integrated pest control and to a great hue and cry for the greater utilization of the technique of biological control. Although the field of weed control has traditionally used integrated control procedures there is no doubt that new advances can and should be made in this area. This has nothing to do with the matter of using more or less herbicide but only with the question of having and using the most suitable techniques considering all factors involved--environmental, safety, and otherwise.

There appears to be the common misunderstanding that integrated pest control can only proceed if we have a massive national force of pest control specialists who can specify and regulate on a field-by-field basis the specific control procedures which are proper and permissible. While there is a need to coordinate to some extent the actions being taken on all classes of pests, the greatest advances in integrated pest control will be made by having the entomologists perfect integrated insect control, the weed scientists perfect integrated weed control, and so on.

There is also a common misunderstanding about the extent to which biological control can apply to the field of weed control. This technique is applicable when a single major weed species occurs on a wide-scale basis and cyclic resurgence of the weed can be tolerated to some extent. A few examples of notable success can be cited for this technique under these conditions and no doubt more progress can be made in the future. It is obvious that the biological technique is not applicable to mixed weed species which are widespread in our croplands and in many other areas where weeds must be managed.

It is abundantly clear that we do not need a high priority national strategy on discovering and developing herbicides which are active on a single weed species. Although some herbicides are notable for their effectiveness on specific weeds and additional such materials of this type could be used, their discovery and development can come through the normal channels which have continued to bring us an abundant variety of unique, useful, and safe herbicides. What is much more necessary is that we have a continuing supply of new herbicides with uniquely useful attributes. Especially desired are those types which are broad in spectrum of control but which can be used safely to produce crops by virtue of crop tolerance conferred either by biological specificity or contrived techniques of use.

It is highly important that we have a national strategy on weeds which involves research on understanding weeds and their impact, research on developing the most effective methods of control, and educational techniques to insure adoption of the preferred techniques. In this regard it is a simple fact that the research laboratories of the world continue to produce a variety of interesting and potentially useful herbicides. Many of these have already met the standards required to permit their use and, hopefully, additional ones will continue to do so. The avoidance of unnecessary and unreasonable regulations can help to assure that this flow of new products will continue. With all of these basic components and others of the mechanical and management types available, it is important that we have weed scientists addressing themselves to organizing the available components into the most effective systems possible.

Even when the weed scientist has perfected the most suitable system feasible there is now and possibly in the future a substantial bar to the utilization of the fruits of his labor. This bar arises because each new use of a given herbicide requires a substantial amount of work to result in the issuance of a label. Even though a given herbicide is labeled for widescale use in many crops and is known to be a safe product, its extension to a new situation defined as being meritorious by a weed scientist has an inadequate chance of being realized. This is likely to be a growing problem in the future and it is one in which state governments, the federal government, weed scientists, commercial firms, and producers have a vested interest and should take action.

In summary, it appears that an appropriate national strategy for the development of selective herbicides for the future should embrace, on the one hand, the identification of a series of principles based on weed science as distinct from those of insect science and, on the other hand, the identification of a series of working guidelines which will encourage and permit the fullest development of present and future technologies.

Chapter 3

BIOLOGICAL AND BIOCHEMICAL CONSIDERATIONS IN THE USE OF SELECTIVE AND NONSELECTIVE FUNGICIDES

Malcolm R. Siegel
Department of Plant Pathology
University of Kentucky
Lexington, Kentucky

Microorganisms which produce plant diseases have been responsible for considerable crop destruction. According to Large {1958} one of the most destructive plant disease outbreaks in the recent past occurred in 1845 and 1846 when the late blight fungus (*Phytophthora infestans*) destroyed the potatoe crop in Ireland. This event resulted in the death of 1 million people and the emigration of another 1-1/2 million. While most plant diseases do not affect man in such a spectacular manner, history is replete with examples of plant diseases which reduced crop yields and thereby affected the economic and social development of countries {Large, 1958; Carefoot and Sprott, 1967}.

Today it is estimated that plant diseases cause an annual 10 percent loss of the U. S. crop {LeClerg, 1964} and an 18 to 20 percent reduction in world agricultural yields {Bent, 1969}. Occasionally greater losses are experienced, as with the coffee bean disease (*Colletotrichum coffeanum*) in Africa in 1967

(33 percent loss) {Bent, 1969} and with corn leaf blight (*Helminthosporium maydis*) in the United States in 1970 (15 percent loss) {Hooker, 1972}.

A variety of methods are available to check or prevent diseases. The breeding of resistant varieties is the primary means of controlling virus diseases and certain fungal and bacterial pathogens. Modification of cultural practices is also of some value. However, the primary method available for continuous control of many plant diseases is by the use of chemicals.

Chemicals were not used for disease control with any regularity until 1882 when Millardet discovered Bordeaux mixture. Indeed it would appear that before fungicides could be developed the germ theory of diseases had to be expanded as by Pasteur in 1860 and the nature of fungus diseases established through the work of Debary and Kuhn, 1853--1863 {Large, 1958}.

Chemical compounds may act as protectants and/or eradicants in controlling plant pathogens. Protection and eradication may occur either on the surface of the plant or plant parts, in the soil or within the plant itself. Those compounds which either remain on the surface of the plant or plant parts and kill or inhibit the pathogen before it enters the plant are considered to be seed, foliar, and fruit protectants (surface protectants). Those compounds which are used to treat the soil and thus protect the plants (which may not be present at the time of treatment) from soil pathogens, are soil protectants. Those compounds which either topically permeate various plant parts, or which permeate and are then translocated within the whole plant and kill the pathogen as it enters, or which rid the plant of an established pathogen, can be designated as systemics. Implicit with their usage is the concept that degrees of biological selectivity exist and that for a chemical agent to be effective it must kill the pathogen without adversely affecting the host plant.

SURFACE PROTECTANT FUNGICIDES

Certain concepts of plant protection through the use of chemical agents were actually postulated by H. Marshall Ward (1880--1881) 2 years before the discovery of Bordeaux mixture {Large, 1958}. The theory, with minor modifications, presupposes that fungicides:

1. Must be applied uniformly over the entire surface of the susceptible tissue in a thin film.
2. Protect without injuring the tissue.
3. Should be designed to rigidly defined solubility and adhere tenaciously to plant surfaces without being either eroded or decomposed by wind, rain, sunlight, or plant secretions.
4. Should be released in concentrations lethal to the pathogen during the infection period and be capable of reaching site(s) in the fungal cell without extensive detoxification.

These properties best characterize the seed, foliar, and fruit protectant fungicides commonly used in plant disease control. From 1882 to 1932 these were either mercury, sulfur, and copper compounds, or combinations of copper and sulfur compounds. After 1934, with the introduction of the dithiocarbamates, they were organic molecules. Organic molecules offer a major advantage over the inorganics in that they can be easily modified. This affords a greater control over the biological selectivity of the compound.

Some common commercial protectant fungicides are listed in Table 1. The most widely used, on a worldwide basis, are the various copper compounds, sulfur, captan and folpet, and the dithiocarbamates.

Selectivity has played an important role in the development of the protectant fungicides. Selectivity of protectants can be discussed from the standpoint of the pathogen, the host plant and the nontarget species.

TABLE 1

Some Common Commercial Surface Protectants

Coppers	Dicloran
Sulfur	Chlorothalonil
Captan-folpet	Drazoxolon
PCNB	Dinocap
Dithiocarbamates	Dichlone
Maneb-Zineb	
Thiram	Dodine
Ferbam	
	Dyrene
Fentin	

Pathogen and Host Plant Selectivity

From the pathogen standpoint, the classical commercially usable seed, foliar, and fruit protectant fungicides are considered to be nonselective. These fungicides are biologically broad-spectrum; they are toxic to many species of fungi in the phycomycetes, ascomycetes, basidiomycetes, and fungi imperfecti. However, this does not mean that all protectants are used commercially to control most fungal pathogens. Commercial usage is dependent on a number of factors other than fungitoxicity as determined in laboratory screening tests. Cost, formulation, and disease situation are important factors which control their usage.

The use of protectant fungicides usually does not depend on their having a specific action against fungi but rather on their exhibiting a differential toxicity, so that a specific dose will kill the pathogen and not injure the host plant. Differential toxicity is based primarily on differences in permeability and accumulation of toxicant in fungi and host plant tissue. This concept of differential toxicity can be illustrated by the following examples:

1. That susceptible fungi can rapidly (1.5-10 minutes) accumulate large toxic concentrations (2.6-24 mg/g dry wt) of various protectant fungicides is illustrated by the data in Table 2. Generally, uptake is either proportional to the external dosage, or to the log of the external dosage {McCallan et al., 1959}. Accumulation of fungicide apparently depends on the availability and number of the reacting sites in the cell.
2. On the other hand, protectant fungicides generally do not permeate and accumulate in plant cells. When Wellman and McCallan {1946} investigated a series of 1-hydroxyethylene-glyoxalidines they determined that the number of carbon atoms of the side chain, attached to the 2-imidazoline nucleus, controlled the permeability and toxicity of the compounds to both the plant and fungal organisms. When the number of carbon atoms was increased to 11, permeation into the foliage, as well as phytotoxicity, reached a maximum. As the hydrocarbon chain was increased to 17 carbon atoms phytotoxicity decreased, while fungitoxicity increased.

TABLE 2

Uptake of Protectant Fungicides at Fungicidal ED50

	mg/g dry wt		
Compound	*Saccharomyces pastorianus*	*Alternaria tenuis*	*Neurospora crassa*
Captan[a,b]	3.8	--	6.9
Copper[a]	--	12	2.6
Dodine[a]	--	24	10
Chlorothalonil[c]	3.2	--	--

[a]{Somers, 1969}.
[b]{Siegel and Sisler, 1968}.
[c]{Tillman et al., 1973}.

Maximum fungitoxicity and minimum phytotoxicity occur for the 17 carbon compound which is the commercial fungicide glyodin.

3. Solel and Edgington {1973} have recently reported that certain protectant fungicides such as captafol, captan, and chlorothalonil show moderate transculticular movement. However, no translaminar movement could be demonstrated. This data suggests that these fungicides will move into the cuticle because of their lipophillic characteristics, but that they will not move from the cuticle into the living tissue of the leaf.

The nonselectivity of most protectant fungicides can be explained by their biochemical fungitoxic modes of action. Lukens {1969} considers most of the protectant fungicides to be general toxicants. They are characterized by rapid uptake, accumulation of fungicide in fungal cells, and by indiscriminate reactions with cellular constituents. In the biochemical sense they are also nonspecific or nonselective in their reactions. In plant and animals cells these compounds may be toxic if they permeate and can be accumulated without being detoxified. This implies that the same sites that are reactive in fungal cells are potentially reactive in plants, as well as in other nontarget species.

The mechanism of action of many of the surface protectants, as general toxicants, involves the inhibition of multienzyme systems required for energy production in the cell {Kaars Sijpesteijn, 1970}. This includes such compounds as the dialkydithiocarbamates (thiram and dimethyldithiocarbamate), 8-hydroxyquinoline (oxine) and triphenyltin (fentin) which react through chelation mechanisms with metal-sensitive enzymes {Kaars-Sijpesteijn, 1970; Lukens, 1971}. Fungicides which apparently operate by nucleophilic substitution reactions include dyrene, dichlone, chloronil, captan and folpet, monoalkyldithiocarbamates (maneb, nabam, and zineb), and chlorothalonil {Lukens, 1971}. Many of these compounds act as biological alkylating agents. Cellular sulfhydryl groups are particularly reactive with this group of fungicides.

The biochemical nature of the multireaction sequence of most of the protectant fungicides can be illustrated with chlorothalonil. The products of the reaction of chlorothalonil with cellular reduced glutathione (GSH) and protein sulfhydryls (Pr-SH) is shown in Fig. 1 {Tillman et al., 1973; Vincent and Sisler, 1968}. Accumulation of large concentrations of the fungicide can be explained primarily on its reactions with GSH. The substitution proceeds as an S_N2 reaction. The mercaptide ion (GS-) attacks the halogen atoms at either the 4 or 6 position of the benzene ring. This is facilitated by the ability of the 1,3 dicyano groups to direct ortho/para substitutions {Vincent and Sisler, 1968}. The reaction sequence yields chlorine-substituted chlorothalonil-glutathione derivatives (Fig. 1, I and II). At the same time chlorothalonil will react with

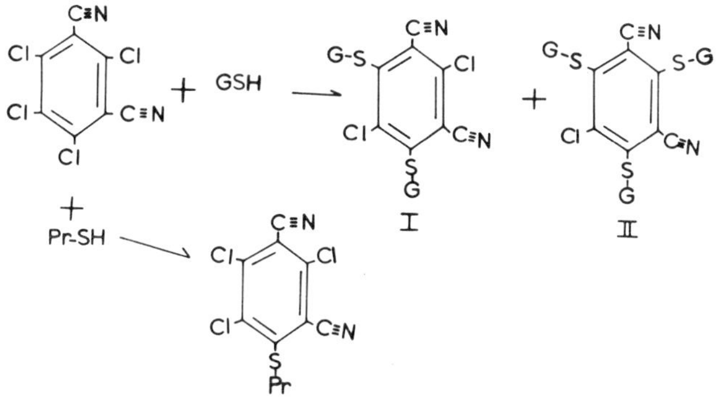

FIG 1. Products of the reaction of chlorothalonil with cellular reduced glutathione (GSH) and protein sulfhydryls (Pr-SH). (I and II, chlorine subsituted chlorothalonil-glutathione derivatives).

protein thiols. This was determined by binding studies using fungicide and enzyme inhibition studies {Tillman et al., 1973; Long and Siegel, 1974}.

In fungal cells all the GSH was reacted with chlorothalonil. This is basically a detoxification reaction. Decreased cell viability and increased inhibition of the enzyme glyceraldehyde phosphate dehydrogenase (E.C. 1.2.1.12) occurred only after the loss of all the GSH. This suggests that the low molecular weight thiol was protecting the enzyme sulfhydryl sites. Irreversible loss of this enzyme and others used in energy production results in toxicity {McCallan et al., 1959; Vincent and Sisler, 1968; Long and Siegel, 1975}.

Because surface protectants cause a multiplicity of reactions in fungal cells, it would be difficult for resistance, based on genetic changes at the various sites of action, to occur. Indeed resistance to most protectant fungicides is rare. Tolerance and resistance, when they do occur, may be based on changes in permeability or on increased detoxification by the organism {Dekker, 1969}.

Nontarget Selectivity

Generally the surface protectants have high nontarget selectivity. When toxicity to nontarget species occurs, it is probably based on the same multisite mode of action involved in fungitoxicity. That is, for toxicity to occur, the protectants must permeate and be accumulated in cells without excessive detoxification. In larger animals excretory processes and detoxification reactions appear to be the primary protective mechanisms. Detoxification involves not only reactions with excess cellular sites, but also metabolic conversion of fungicides to nonfungitoxic compounds. In lower plant and animal species the primary protective mechanisms probably involve the relative inability of these compounds to enter the cells. The summary data in

TABLE 3

Effect of Some Protectant Fungicides on Nontarget Species[a,b]

Chemical	LD50 (mg/kg:rat)	Nontarget toxicity
Coppers	100--300	Fish, soil, animals
Mercuries	50--250	Accumulation in animals, birds
Captan	9,000	Weak mutagen and teratogen,[c,d] fish fry
PCNB	12,000	--
Maneb	7,500	Ethylenethiourea (carcinogen[e]
Ferbam	17,000	Fish, phytoplankton, insect predators
Chlorothalonil	10,000	Fish fry
Dyrene	2,710	Phytoplankton
Dichlone	1,300	Insect predators, fish
Fentin	108	Fish

[a]{Pimental, 1971}.
[b]{British Crop Protection Council, 1968}.
[c]{Collins, 1972}.
[d]{Shea, 1972}.
[e]{Graham et al., 1973}.

Table 3 indicate that only the metal-containing fungicides (coppers, mercuries, and tin) have low enough oral LD50 values (100--300 mg/kg) to be considered to have any potential mammalian toxicity. These values, however, are considerably higher (10--100 times) than some of the potentially hazardous insecticides {Metcalf, 1972}.

Some nontarget organisms are quite sensitive to the surface protectants (Table 3). Fish fry of certain species are particularly sensitive to the chlorine-containing fungicides (captan, folpet, dyrene, and chlorothalonil).

Organic mercury compounds are no longer used in the United States because mercury accumulation in plants and animal species adds to the widespread environmental contamination by these compounds.

The captan-folpet group of fungicides has been reported to be weak mutagens and teratogens {Collins, 1972; Shea, 1972}. Ethylenethiourea, a decomposition product and contaminant of the ethylenebisdithiocarbamate fungicides, produced thyroid carcinomas in rats after feeding 250--500 mg/day for 12 months {Graham et al., 1973}.

SOIL PROTECTANTS AND DISINFECTANTS

Chemicals used to control soil-borne pathogens can be divided into two groups. The first group of chemicals acts as a protectant in the manner previously described. That is, the same biological and biochemical considerations are applicable as were described for the surface protectants. The chemicals can be applied directly to the soil as either granular or liquid formulations. The plants may be present during treatment or can be planted immediately after application. Compounds which can be used in this manner include the dithiocarbamates, captan, PCNB, ethazol, diazoben, etc.

The second group of compounds are volatile disinfectants and are toxicants which control fungi, bacteria, nematodes, weeds, and insects. Some are general biocides and are nonselective for target and nontarget species in the soils. Plants are not present at the time of treatment and may not be planted for a period of time, depending on the chemical used. Fumigants are formulated as liquids, granules, and gases and include such compounds as: methyl bromide, chloropicrin, vapam, mylone, formaldehyde, nemagon, D-D, vorlex, etc.

The mechanism of action of the fumigants is based on the same principles as described for the surface protectants {Lukens,

1971}. They are multisite reactors (general toxicants) and the compounds therefore must accumulate in cells. Their nonselectivity towards nontarget species may be based on a lack of permeability barriers to the vapors. The use of most fumigants is generally restricted to greenhouse and seed beds since various seals must be used to keep the material in the soil. However, certain compounds which release their vapors slowly can have more general usage.

SYSTEMIC CHEMOTHERAPEUTANTS

During the last 10 years a new group of chemicals has been introduced commercially. They are selective enough to allow for a differential toxicity between plant and pathogen (large therapeutic index) if the fungicide is within the plant.

The history and concepts of the chemotherapy of plant diseases by employing chemicals which move within the plant (systemic) have been recently reviewed by Wain and Carter {1971}.

Plant chemotherapy may be defined as the control of plant diseases by compounds that, through their effect upon host or pathogen, reduce or nullify the adverse effects of the pathogen after it has entered the plant. Systemic chemicals acting as chemotherapeutic agents must move within the plant where they either kill the pathogen as it enters or rid the host of an established pathogen. These effects may be accomplished either directly, by the chemical itself, or indirectly, by inducing the plant to attack the pathogen.

Systemics should have the following characteristics:

1. They must easily permeate plant cells without injury.
2. They must be translocated in either the xylem or phloem and be uniformly distributed in the plant.
3. They must resist detoxification by the plant for a reasonable time.

4. They must compete with the established protectant fungicides from the standpoint of cost and application.

Pathogen and Host Selectivity of Systemic Fungicides

The commercial systemic chemicals listed in Table 4 show varying degrees of pathogen selectivity almost exclusively within the fungi. The only antibacterial agents are the antibiotics streptomycin and kasugamycin. Except for chloroneb and one or two of the antibiotics, none of the compounds listed are effective against the phycomycetes. The activity of the oxathiins, pyrimidines and morpholines is restricted to certain genera in the basidiomycetes and actinomycetes. The latter two groups of chemicals are narrowly pathogen-selective within the ascomycetes to the family Erysiphales (powdery mildews). The antibiotics show narrow pathogen-selectivity to various fungal and bacterial organisms. The most wide-spectrum are the benzimidazoles and thiophanates whose breakdown product (methyl 2-benzimidazolecarbamate, MBC) is the primary fungitoxic principle for both chemical groups {Kaars Sijpesteijn, 1972}.

All the recently introduced synthetic systemic compounds listed in Table 4 move unidirectionally in the transpiration stream within the xylem tissue. These compounds are considered to move in the apoplast (nonliving tissue) as distinguished from the symplast (living tissue) {Evans, 1971; Crowdy, 1972}. Bipolar movement of a significant amount of systemics in the phloem system has not as yet been demonstrated. Once these apoplastic compounds enter the leaf they will not be redistributed to other portions of the plant but tend to move acropetally toward the leaf margins. When systemics move in the xylem, transpiration determines the rate and amount of movement of fungicide. The concentration of the fungicide is therefore dependent on the ability of the various plant organs to transpire {Evans, 1971; Crowdy, 1972}.

TABLE 4

Some Commercial Systemics[a,b]

Chemical group	Representative analogue	Pathogen selectivity[c]
Oxathiin	Carboxin	B
Benzimidazole	Thiabendazole	
	Benomyl	A, B
Pyrimidine	Ethirimol	A (Erysiphales)
Thiophanate	Thiophanate	A, B
Morpholine	Tridemorph	A (Erysiphales)
Piperazine	Triforine	A, B
Pyrazolopyrimidine	Pyrazophos	A
Phosphorothiolate	Kitazin	A
Phenol	Chloroneb	P, A, B
Antibiotics (misc.)	Blasticidin S	Bacterial, fungal
	Cycloheximide	Fungal
	Streptomycin	Bacterial, fungal
	Polyoxin D	Fungal
	Kasugamycin	Bacterial, fungal

[a]{Evans, 1971}.
[b]{Woodcock, 1972}.
[c] A = ascomycetes; B = basidiomycetes; P = phycomycetes.

The phytotoxicity of these compounds varies. Some of the antibiotics clearly show phytotoxicity to plants and this has been one of the limiting factors in their use. Although the others, listed in Table 4, are considerably less toxic to plants, some phytotoxicity still exists. For example some phytotoxicity by benomyl can be seen at the leaf margins where excess accumulation of the fungicide occurs {Biehn and Dimond, 1970}.

With the systemics, the mechanism of differential toxicity to pathogens and host plants is rather more complex than that reported for the protectant fungicides. As will be described in the next section of this chapter, the systemics do have a rather specific or selective mechanism of action against the pathogen. Therefore, a lack of toxicity of a systemic to plant cells can be based either on a loss of a reaction site, or a reduced affinity for a reaction site. Another likely possibility is that since all the current agricultural systemics move in the apoplast, accumulation of significant concentrations of toxicant in the symplast, where inhibition of metabolic processes could occur, is limited. There is data to support this hypothesis for certain systemics. Solel et al. {1973} have recently demonstrated that uptake of MBC by the cytoplasm of mesophyll cells occurs, but that it is of a low order. Translocation of MBC from leaf surfaces, via the phloem, to nontreated portions of plants was also demonstrated but only at high application dosages. The persistance of MBC in tissue of various plant species also suggests that the compound is only slowly released into living cells where it can be metabolized {Solel et al., 1973; Siegel, 1973}. A concentration gradient of MBC could exist between living and nonliving cells, with the equilibrium primarily in the direction of nonliving tissue. However, phytotoxicity at leaf margins also suggests that metabolic processes can be inhibited if enough of the toxicant can be accumulated in living tissue.

The agricultural systemics are considered specific toxicants as defined by Lukens {1969}. These compounds interact with either a single or a few cellular sites. This results in the direct inhibition of one or two metabolic cellular functions. Because only few sites are involved, these compounds do not generally accumulate in fungal cells. This type of inhibition, involving possibly one metabolic function, may also, in part, control the degree of pathogen and nontarget selectivity and the development of pathogen resistance which is common with the specific-site inhibitors. It should be mentioned that while all the systemics appear to be specific-site inhibitors, not all specific-site inhibitors are systemics. Dicloran, a protectant fungicide, is reported to be in this class of toxicants {Luken, 1971}.

The known cellular sites and mechanisms of action for some systemics are listed in Table 5. While it is impossible to give details on the mechanism of action of all these compounds, certain mechanisms will be discussed to illustrate various aspects of the selective action of these compounds.

The data in Table 6 indicate the magnitude of the differences in uptake of fungicide between two systemic specific-site inhibitors and captan, a protectant and multisite inhibitor. The differences in uptake are not only based on fewer reactions with sites involved in toxicity, but also on a lack of detoxification of the two systemics in the treated cells. Fungal cells can accumulate specific-site inhibitors when either detoxification reactions occur or when more numerous metabolic functions are inhibited. This may be occurring in chloroneb-treated cells which accumulate up to 2,000 μg/gm wet weight of cells {Hock and Sisler, 1969}. Chloroneb inhibits the process of cell wall formation in an unknown manner {Hock and Sisler, 1969}.

Pathogen resistance, as well as tolerance and resistance of host plants and nontarget species to the systemics can be grouped according to two different mechanisms {Dekker, 1972}. In the first group there is resistance and selectivity based on either permeability or detoxification mechanisms. In the second group the mechanisms involve the reaction site. These include decreased affinity of the toxicant for the site, lack of the site, and a circumvention (bypass) of the site. It appears that pathogen resistance to many of the systemics which are listed in Table 5 involves site mechanisms. However, there are exceptions such as resistance to blasticidin {Dekker, 1972} which involves permeability changes in resistant mutants.

TABLE 5

Sites and Mechanisms of Action of Systemics

Site	Interference with (mechanisms)	Chemical
Cell wall	Chitin synthesis[a]	Kitazin Polyoxin Hinosan
Protoplast-membrane	Permeability[b]	Nystatin
Mitochondria	Energy production	Carboxin[c,d] Benomyl[e] Thiabendazole[f]
Endoplasm reticulum	Sterol synthesis[g]	Triforine
Ribosomes	Protein synthesis[b]	Blasticidin Cycloheximide Streptomycin Kasugamycin
Nucleus	Mitosis[h,e,i]	MBC[j] (Benomyl) MBC (Thiophanate) Thiabendazole

[a] {Kaars Sijpesteijn, 1972}.
[b] {Munoz et al., 1972}.
[c] {Ulrich and Mathre, 1972}.
[d] {Lyr et al., 1972}.
[e] {Hammerschlag and Sisler, 1973}.
[f] {Allen and Gottlieb, 1970}.
[g] {Sherald et al., 1973}.
[h] {Evans, 1971}.
[i] {Davidase, 1973}.
[j] MBC (Methyl 2-benzimidazolecarbamate)

Nontarget Selectivity

Lack of toxicity of the systemics to the host plant and non-target species can be based on both site and nonsite mechanisms. The concept of differential toxicity of the systemics toward the host plant was previously discussed from the standpoint of the lack of permeability and accumulation between the apoplast and

TABLE 6

Uptake of Selective Fungicides

	μg/g wet wt	
Compound	*Saccharomyces pastorianus*	*Neurospora crassa*
Benomyl[a]	--	1.4
Cycloheximide[b]	0.38	--
Captan[c]	11,400	20,700

[a]{Clemons and Sisler, 1971}.
[b]{Wescott and Sisler, 1964}.
[c] Conversion of data in Table 2 from dry wt. to wet wt. was achieved by arbitrarily assigning a value of 75 percent water content in the cells of *S. pastorianus* and *N. crassa*.

the symplast. However, mechanisms involving the site can also be demonstrated and used to explain the high degree of selectivity of the systemics. The compounds Kitazin, Hinosan, and Polyoxin affect cell wall formation by inhibiting chitin synthesis in the pathogen. Since this metabolic process is not present in either plants or animals these compounds should exhibit high selectivity. When similar metabolic processes are present in target as well as nontarget species the systemic may have an affinity for only the site in the pathogen. For example, carboxin significantly (>50 percent) inhibited mitochondrial bound succinic dehydrogenase in susceptible species at concentrations as low as 0.1 μM. It required up to 1,000 times as much fungicide (100 μM) to inhibit succinic dehydrogenase in tolerant and resistant fungal, plant, and animal species {Ulrich and Mathre, 1972}.

The systemics listed in Table 7 do not pose any great mammalian hazards. Except for the organophosphate compounds triamiphos, pyrazophos, and kitazin and the antibiotics

TABLE 7

Acute Mammalian Toxicity of Some Commercial Systemics[a]

Chemical	Rate LD50 mg/kg
Triamiphos	20
Blasticidin	39
Cycloheximide	133
Pyrazophos	140
Kitazin	238
Tridemorph	1,270
Polyoxin	1,500
Carboxin	3,200
Thiabendazole	3,220
Ethirimol	4,000
Triforine	6,000
Streptomycin	9,000
Benomyl	9,590
Chloroneb	11,000

[a]{Woodcock, 1972}.

blasticidin and cycloheximide, all systemics listed have oral LD50 values for rats of over 1,000 mg/kg. In this respect they resemble the relatively safe surface protectants. While it is difficult to compile a list of effects of systemics on nontarget species, this does not mean that such effects do not occur. For example, the development of triarimol, a promising systemic, has been suspended because of undesirable toxicological effects. Benomyl, a widely used compound, has been shown to be a miticide {Delp and Klopping, 1968} and to be toxic to earthworms {Stringer and Wright, 1973}. Only with continued use of the systemics will it become possible to determine their complete effects on nontarget species. It is probable that the more narrowly pathogen-selective systemic compounds will be found to be less toxic or toxic to fewer nontarget species than the more broad-spectrum protectants.

Unlike the protectants, which can be removed from the surface tissue, the systemics are intended to be retained within the plant. Therefore, not only the toxicology of the parent compound, but also the various degradation products produced by plant metabolism are of primary importance and must be investigated.

Usage

Selective systemics have found considerable use in controlling diseases that are not easily controlled with the protectants. Even when protective fungicides control a disease, systemics are often used because of the advantages offered by their unique action. The use of systemics recently has been reviewed by various authors in the book Systemic Fungicides {Marsh, 1972} and can be briefly summarized.

There has been considerable success in the control and eradication of pathogens in plant propagules (seed, seed pieces, bulbs, etc.) and in fruits and vegetables (post harvest control). The treatment of plant propagules in certain cases has led to the control of airborne pathogens. This occurs because the chemical dressing on the propagules is acting as a reservoir of fungicide during the growth of the plant.

In traditional soil treatment, fungicides were added to soil to kill soil-borne pathogens prior to planting or placed near the germinating plants to act as protectants. Systemics have also been used for this purpose. Because of their movement in the xylem (upwards and outwards), the control of airborne pathogens of annual and perennial crops can theoretically be achieved, following soil treatment and root uptake. However, in the field, the control of soil-borne and airborne pathogens has proven to be the exception rather than the rule. Where fungi are dispersed throughout the soil the protection of extensive root systems is particularly difficult and costly. If the chemical is poorly distributed, because of soil-chemical interactions, then uneven control of airborne pathogens due to marginal root uptake will occur. Where the soil, roots, and chemical can be contained within either a bed or pot, the control of soil-borne and airborne pathogens is excellent.

Foliar sprays of systemics have been shown to be either equal to or more effective than protectant sprays for control of specific airborne pathogens. Stem treatments (injection) of woody plants also show promise in control of vascular diseases such as Dutch elm. However, it is questionable whether foliar applications of systemics which move only in the xylem should be used when there are protectants which give good control of a disease. Foliar sprays do not take complete advantage of the systemicity of the compounds. Since all are translocated apoplastically, movement into the new or actively growing tissue is limited and is probably dependent on the repeated application of material to nonfoliar above-ground portions of the plants. Even though fewer applications of the systemics are needed, the number may still be great enough to result in selection pressure on the fungal population, thus hastening the development of resistant pathogens.

When systemics which move in the symplast are developed, the advantages of foliar application will be greater. Since these compounds will move to, and be accumulated in the roots, the control of soil-borne diseases will be facilitated.

SUMMARY AND CONCLUSIONS

Chemicals used to control plant diseases have been discussed from the standpoint of their usage and their biological and biochemical selectivity. These agricultural chemicals were divided into two broad groups, designated as protectants and systemics. Protectants were shown to be biologically pathogen-nonselective, and almost all act as general toxicants biochemically. Since these protectants are nonselective in their modes of action it might be expected that they would not only affect the target but also nontarget species. Except for the soil fumigants and sterilants they appear to be generally selective for nontarget species. Some phytotoxicity and nontarget effects were, however, demonstrated for this group of chemical control agents. The differences in toxicity between target (fungi) and nontarget organisms appear to reside primarily in permeability and accumulation phenomena and detoxification mechanisms. Because protectants react with a multiplicity of reaction sites in fungal cells, resistance based on genetic changes at these sites is rare.

Systemics were shown to be biologically pathogen-selective and biochemically specific-site inhibitors. Because of their specific or selective modes of action, systemics should and do show high selectivity towards nontarget species. This does not imply that reaction site(s) are not present in nontarget species. Phytotoxicity and other nontarget affects appear to be minimal. Because all the commercial systemics move within the apoplast, differential toxicity could involve the inability of the compound to permeate and accumulate within the symplast, where plant metabolic functions occur. On the other hand, tolerance and resistance in fungi, host plants, and nontarget species may also involve factors other than permeability. Lack of a site or a change in the affinity for the site of action may also occur.

Pathogen resistance to systemics appears to be quite common {Dekker, 1972}. While various mechanisms of resistance are possible, one might predict that a mechanism involving the specific reaction site predominates.

If one assumes that the surface protectants and systemics have been developed to have minimal nontarget effects, then their use will depend upon factors other than safety. Among these include development costs, screening systems, usage (disease situation), and the development of pathogen resistance.

In 1965 it was estimated that it cost approximately $2.92 million to develop a pesticide {Wellman, 1967}. This value has tripled, and increased costs have not been due solely to inflation. The Federal Environmental Pesticide Control Act of 1972 requires necessary but costly and time-consuming research into the effectiveness, usage (labeling), residues, and toxicology of the pesticide and major breakdown products. If the product is marketable it may require 6--10 years for a company to generate adequate profit margins. While it is not the purpose of this paper to discuss product developments and costs, they do have an effect on new fungicide development. Since organic fungicide sales in 1971 in the United States represented only 7.5 percent of the total organic pesticide sales value {USDA, 1972}, it becomes rather obvious that costs and profit margin are of prime importance in the development of these chemicals. This may be reflected in the fact that more than 50 percent of the newer systemics have been developed in Japan or Western Europe. While this may be due to the need for control of rather specialized disease situations (i.e., rice blast in Japan) it may also reflect lower development costs and less stringent legal requirements.

An important feature in the development of new selective chemicals will be the availability of adequate screening procedures. For example, synthetic chemicals which move in the plant and are either activated into toxic moieties by the host or pathogen, or stimulate the host to produce a toxic compound, may require novel screening procedures for detection.

As was previously discussed, selective systemics have potential advantages over the nonselective protectants in their usage. These advantages involve their novel selectivity, penetration, and translocative properties. Fewer applications at lower rates is an attractive feature for the user. To the environmentalist, their selective nature should also be attractive (assuming high nontarget selectivity). From the plant pathologists' view, selective systemics offer opportunities to control diseases that either have not been controlled before by chemicals, or which it was not economically feasible to control in the past.

The major disadvantage, and the factor which may limit the long-term usage of a given selective systemic or protectant, is the development of resistant pathogens {Evans, 1971}. There are certain procedures which might reduce the selection pressure on a given pathogen population and hence the onset of resistance. These include:

1. Following mixed treatment schedules or using mixtures of selective and nonselective compounds.
2. Reducing the period of contact with the pathogen (fewer applications) and avoiding sublethal dosages wherever possible. Once resistance has been established, there is no alternative but to use chemicals with a different mode of action, if they are available.
3. Developing systemic chemicals which are activated to toxic forms only by the pathogen and act as general toxicants. Activation would occur only in the few plant cells involved in pathogenesis and would result in death to both the pathogen and host cells.

The use of either mixed chemical applications or alternating selective and nonselective chemicals appears to offer the best approach to the resistance problem for the immediate future. For this reason and also because of development costs and disease situations where only one type of chemical is available, it will be necessary to use both the selective systemics and nonselective protectants in plant disease control.

ACKNOWLEDGMENTS

Published with the permission of the Director of the University of Kentucky, Agricultural Experiment Station as Journal article No. 73-11-165. Supported in part by Public Health Service Grant ES 00227 from the National Institute of Environmental Health Sciences.

REFERENCES

Allen, P. M., and D. Gottlieb, 1970. Appl. Microbiol. *20*:919.

Bent, K. J., 1969. Endeavour. 28:129.

Biehn, W. L., and A. E. Dimond, 1970. Plant Dis. Reptr. *54*:12.

British Crop Protection Council, 1968, *Pesticide Manual*, (A. Martin, ed.), Worchester, England.

Carefoot, G. L., and E. R. Sprott, 1967. *Famine on the Wind*, Rand McNally & Co., New York.

Clemons, G. P., and H. D. Sisler, 1971. Pest. Biochem. Physiol. *1*:32.

Collins, T. F. X., 1972. Fd. Cosmet. Toxicol. *10*:353.

Crowdy, S. H., 1972. *Systemic Fungicides*, (R. W. Marsh, ed.), John Wiley & Sons, New York, p. 92.

Davidase, L. C., 1973. Pest. Biochem. Physiol. *3*:317.

Dekker, J., 1969. World Rev. Pest Control. *8*:79.

Dekker, J., 1972. *Systemic Fungicides*, (R. W. Marsh, ed.), John Wiley & Sons, New York, p. 156.

Delp, C. J., and H. I. Klopping, 1968. Pl. Dis. Reptr. *52*:95.

Evans, E., 1971. Pestic. Sci. *2*:192.

Graham, S. L., W. H. Hansen, K. J. Davis, and C. H. Perry, 1973. J. Agr. Food Chem. *21*:324.

Hammerschlag, R. S., and H. D. Sisler, 1973. Pest. Biochem. Physiol. *3*:42.

Hock, W. K., and H. D. Sisler, 1969. Phytopathology. *59*:627.

Hooker, A. L., 1972. J. Environ. Quality. *1*:244.

Kaars Sijpesteijn, A., 1970. World Rev. Pest Control. *9*:85.

Kaars Sijpesteijn, A., 1972. *Systemic Fungicides*, (R. W. Marsh, ed.), John Wiley & Sons, New York, p. 132.

Large, E. C., 1958. *The Advance of the Fungi*, Jonathan Cape, London.

LeClerg, E. L., 1964. Phytopathology. *54*:1309.

Long, J. W., and M. R. Siegel, 1975. Chemico-Biol. Interactions *10*:383.

Lukens, R. J., 1969. *Biodeterioration of Materials*, Elsevier Publishing Co. Ltd., Barking, England, p. 486.

Lukens, R. J., 1971. *Chemistry of Fungicidal Action*, Springer-Verlag, New York.

Lyr, H., G. Ritter, and G. Casperson, 1972. Zeitschrift für Allg. Mikrobiologie. *12*:271.

Marsh, R. W., (ed.), 1972. *Systemic Fungicides*, John Wiley & Sons, New York.

McCallan, S. E. A., H. P. Burchfield, and L. P. Miller, 1959. Phytopathology. *49*:544.

Metcalf, R. L., 1972. *Pest Control Strategies for the Future*, National Academy of Sciences, Washington, D. C., p. 137.

Munoz, E., F. Garcia-Fernandiz, and D. Vazquez, (ed.), 1972. *Molecular Mechanisms of Antibiotic Action on Protein Biosynthesis and Membranes*, Elsevier Scientific Publishing Co., New York.

Pimental, D., 1971. *Ecological Effects of Pesticides on Non-Target Species*, Office of Science and Technology, U. S. Government Printing Office, Washington, D. C.

Shea, K. P., 1972. Environment. *14*:22.

Sherald, J. L., N. N. Ragsdale, and H. D. Sisler, 1973. Pestic. Sci. *4*:719.

Siegel, M. R., 1973. Phytopathology. *63*:890.

Siegel, M. R., and H. D. Sisler, 1968. Phytopathology. *58*:1123.

Solel, Z., and L. V. Edgington, 1973. Phytopathology. *63*:505.

Solel, Z., J. M. Schooley, and L. V. Edgington, 1973. Pestic. Sci. *4*:713.

Somers, E., 1969. World Rev. Pest Control. *8*:95.

Stringer, A., and M. A. Wright, 1973. Pestic. Sci. *4*:165.

Tillman, R. W., M. R. Siegel, and J. W. Long, 1973. Pest. Biochem. Physiol. *3*:160.

Ulrich, J. T., and D. E. Mathre, 1972. Jour. Bact. *110*:628.

U. S. Department of Agriculture, 1972. *The Pesticide Review*, Agricultural Stabilization and Conservation Service, Washington D. C., p. 58.

Vincent, P. G., and H. D. Sisler, 1968. Physiologia Plantarum. *21*:1249.

Wain, R. L., and G. A. Carter, 1972. *Systemic Fungicides*, (R. W. Marsh, ed.), John Wiley & Sons, New York, p. 6.

Wellman, R. H., 1967. *Fungicides*, vol. 1, (D. C. Torgeson, ed.), Academic Press, New York, p. 125.

Wellman, R. H., and S. E. A. McCallan, 1946. Contrib. Boyce Thompson Inst. *14*:151.

Wescott, E. W., and H. D. Sisler, 1964. Phytopathology. *54*:1261.

Woodcock, D., 1972. *Systemic Fungicides*, (R. W. Marsh, ed.), John Wiley & Sons, New York, pp. 34 and 86.

Chapter 4

PRACTICAL CONSIDERATIONS IN USE OF SELECTIVE INSECTICIDES AGAINST MAJOR CROP PESTS

T. F. Watson
Department of Entomology
University of Arizona
Tucson, Arizona

A review of the literature gives the impression that there is wide divergence among researchers on the meaning of selectivity as it relates to pesticides. It therefore seems appropriate to start this paper by defining selectivity. Stern et al. {1959} define a selective insecticide as, "an insecticide which while killing the pest individuals spares much or most of the other fauna, including beneficial species, either through differential toxic action or through the manner in which the insecticide is utilized (formulation, dosage, timing, etc.)." Bartlett {1964} applies essentially the same meaning to the term when he says that, "selectivity is the measure of the capacity of a treatment to spare natural enemies while destroying pests. The term is ordinarily used in a relative sense to express the differential favorability of a natural-enemy-pest ratio evoked by one treatment." Bartlett's definition seems to differ slightly from Stern's in that he emphasizes the natural enemy-pest ratio evoked by a treatment rather than the killing of a pest per se while sparing the natural enemies.

Gasser {1966}, on the other hand, believes that the use of pesticides in a selective manner is subject to two conditions: "(1) the crop goes through periods of vegetation, during which the pests that should be controlled are exclusively present and can be killed by the application of any effective pesticide or, (2) selective preparations are used, which are able to control the given pest or group of pests without harming the rest of the fauna, particularly predators, living on the crop." He states that the first condition rarely occurs and in essence is limiting his concept of selective use of pesticides strictly to differential physiological selectivity.

Metcalf {1966} states that the definition of the word insecticide as "an agent for destroying insects" clearly implies selectivity of action. He further states, however, that many substances used to kill insects are inherently hazardous to higher animals and that their selectivity is apparent only through careful application and confinement to the treated area. He therefore clearly recognizes physical or ecological selectivity and stresses that "much additional research remains to be accomplished in the general area of improving the selectivity of intrinsically non-selective materials by their selective application and utilization." Relative to truly physiologically selective insecticides he states "it seems clear that the chemist and toxicologist are capable of tailoring insecticidal molecules incorporating almost any combination of desirably selective properties."

Smith {1970} states that, "while we await the development of pesticides showing differential physiological effects we should enhance selectivity of available ones by taking advantage of ecological and behavioral characteristics of the pest species as well as the physical characteristics of formulations and application techniques to obtain differentail kill of undesirable organisms (i.e. selective action)."

Many examples have been reported illustrating both physiological selectivity {Stern et al., 1959; Reynolds et al., 1959; Metcalf, 1966; Hall and Dunn, 1959; Hoyt, 1969; Ridgway et al., 1967; MacPhee and Sanford, 1961; Lord, 1947; 1949; Lingren and Ridgway, 1967; Getzin, 1960; Pickett, 1959; and others} and physical selectivity {Stern and van den Bosch, 1959; Smith, 1970; Ridgway et al., 1967; Falcon et al., 1968; Pickett, 1959; Hoyt, 1969; Westigard, 1971}. Both concepts are aimed at the utilization of insecticides in conjunction with beneficial organisms.

In any event, there seems to be general agreement among most who are concerned with pest control that unilateral reliance on chemical pesticides is a thing of the past but that little help can be expected from beneficial organisms when pesticides are present unless "selective" insecticides are used or insecticides are used in a selective manner.

Metcalf et al. {1962} state that, "there is no doubt that the greatest single factor in keeping plant-feeding insects from overwhelming the rest of the world is that they are fed upon by other insects." Newsom {1966} stressed the compatibility of the two when he stated, "applications of insecticides in a discriminatory manner can reduce and in many cases almost eliminate the disruptive effects of even the highly toxic persistent insecticides." Both Newsom {1966} and Smith {1968}, while advocating the integrated control or pest management approach, clearly emphasize the fact that chemical pesticides are, and probably will remain, our main tool for pest control, at least for the foreseeable future.

The search must therefore continue for more selective insecticides. As Metcalf {1966} stated, "it is, however, in the discovery of insecticides that are truly selective in a physiological and biochemical way, that the chemist and toxicologist have the greatest opportunity to produce the genuinely sophisticated and desirable insecticide of the future." At the same time entomologists must learn to use the insecticides currently available to obtain selective action. It is this latter point that we shall consider in the remainder of this chapter.

PEST CONTROL STRATEGY

If we are to make significant advances in pest control we must truly think in terms of pest management. This means utilizing ecologically sound principles to govern pesticide use rather than making generally indiscriminate applications regardless of the nature of the pest problem or the side effects--as has been a common practice during the past 25 years. This necessitates a major change in pest control strategy. Chant {1964} defines strategy as "our general approach to control, including decisions on which method should be used."

In general, the strategy used by most growers since the late 40s--and the one that has been promulgated by entomologists and others--has been almost complete reliance on chemical pesticides for control. Not only that, but the demand for 100 percent mortality has been omnipresent. Both are incompatible with effective strategies of pest management.

One of the major obstacles in shifting emphasis to the integrated control or pest management approach is reorienting the grower to accepting something less than 100 percent mortality of the pest species present, i.e., pest management instead of annihilation. "Presently our use of toxic chemicals is crude and imprecise" {Chant, 1964}. The procedures in the future must be to achieve selective action by using <u>the best selective insecticide available</u>--in terms of environmental compatibility. Van den Bosch and Stern {1962} equated selectivity not only with restricted toxicity but also with precision in timing and the calculation of effective rates of application. This was broadened to include the benefits derived from use of materials with short residual life, methods of application, such as seed treatments, etc., spot treatments restricted to areas where unfavorable pest/beneficial insect ratios exist, or use of any other forms of chemical applications that tend to enhance survival of beneficial species.

CONSIDERATION OF ECONOMIC THRESHOLDS

A number of major considerations affect the proper use of selective insecticides or their use in a selective manner. Probably of greatest significance among these is the utilization of economic thresholds on which to base pesticide applications. On most agricultural crops there are relatively few key pests, i.e., those which are generally present and usually cause economic damage at some time during the growing season. Stern et al. {1959} define economic threshold as, "the density at which control measures should be determined to prevent an increasing pest population from reaching the economic-injury level." This will, of course, vary with different crops and even on the same crop, depending on stage of growth, time of the season, and the complex of beneficial insects associated with the pest. Two main problems are implied in this definition, both of which must be resolved before grower acceptance becomes commonplace. First, the pest population level required to produce economic damage must be established and accepted. This is not an easy task and relatively few sound economic injury levels have been established. Stern {1966} stated that, "in many crops, treatment is often invoked prematurely or unnecessarily because economic thresholds are unrealistically low, poorly defined, erroneously derived, or non-existent." This should not, however, preclude the use of the best information available to set arbitrary levels as an interim basis for initiating control. The very fact that sound economic threshold data are generally lacking has long been used as a basis to initiate prophylactic treatments at the first sign of pest activity. Ecologically, this practice is unsound even if a truly physiologically selective pesticide were available for the pest in question. Any pest undoubtedly represents one component of a food chain which also includes certain predators and parasites. Temporary eradication of the pest would therefore produce adverse effects on the beneficial fauna involved.

The second problem area in the definition, ". . .to prevent an increasing pest population from reaching the economic-injury level," is that of predictability of pest members. Relative to this point, Chant {1964} stated, "Pesticides should be used in corrective rather than preventative programs and this requires us to be able to predict future trends in populations. Moreover, as ability to predict will lead to more sensible pesticide use, the combination of other methods with pesticides will become more feasible." Although considerable progress has been made in recent years on predictability of insect populations by use of mathematical models and computer programming {Watt, 1963}, most pest management programs still rely on periodic field sampling to assess population levels of both destructive and beneficial species and associated crop damage. The utility of these periodic assessment data depends on the knowledge of the population dynamics of the pest species, the suppressive capabilities of the beneficial species, and the capacity of the crop to recover from pest injury.

EVIDENCE OF NEED FOR SELECTIVE INSECTICIDES

Evidence has accumulated over the past 20 years pointing to the need for selective insecticides. Initially, it involved only persons directly concerned with the problem of pest control--the insecticide manufacturer, the grower, the field entomologist.

Adverse side effects arising after the introduction of broad-spectrum insecticides included pesticide-resistant pest strains or species, destruction of nontarget organisms, emergence of previously unimportant species to pest status, resurgence of treated infestations, excessive toxic residues, and legal complications {Stern et al., 1959}. More recently, and perhaps of greater importance to all of us, is the public outcry to ban pesticides. This clamor was spawned partly from the misuse of pesticides but largely from a public poorly informed on the actual hazards involved and without knowledge of the need for pesticides.

Newsom {1966} alluded to some of these problems that were apparent in the early stages of the new synthetic organic pesticide era in a paper on the essential role of chemicals in crop protection. He quoted from the presidential address by Lyle to the American Association of Economic Entomologists that, "We have the technical knowledge and equipment to eradicate the house fly, horn fly, cattle grubs, cattle lice. . ." {Lyle, 1947}. In this address Lyle quoted from a letter from Dr. E. W. Dunnam, U. S. Department of Agriculture, that, "following 5 percent gamma in talc, applied by airplane at rates of 10 to 12 lbs. per acre, the following were taken from the ground dead or paralyzed: aphids, beetles of several species, bollworm moths, bollworm larvae . . .(and others). In a field treated with 12 lbs. per acre of the same material applied 6 times by airplane, no insects were ever seen, although there were a few red spiders on practically every leaf." Newsom stated that Lyle's optimistic views were not surprising but that it was significant that Dunnam's observations on the general presence of spider mites on plants treated with benzene hexachloride did not attract more attention.

Observations and reports became more prevalent on resurgence of treated species or secondary pest outbreaks following the use of pesticides. Ripper {1956} reported that of the 5,000 insect and mite species of economic importance, about 50 species had thus far shown resurgence. Ripper pointed out that this phenomenon was spread over a number of very different groups of phytophagous arthropods and its occurrence followed the application of several types of pesticides with different modes of action. He indicated that chemical control in these cases upset the population dynamics of the organisms involved. He cites three hypotheses that were put forward to explain pest resurgences:

1. The reduction of natural enemies by the pesticide, along with the pest
2. Favorable influences of pesticides on phytophagous arthropods
3. The removal of competitive species

Most explanations offered for pest resurgence or secondary pest outbreaks tend to substantiate the first hypothesis. Much of this is accepted, however, without conclusive data. DeBach's experiments {1946, 1955} demonstrated conclusively how pest outbreaks are caused when he used a selective insecticide in a manner which killed beneficial species but had little if any effect on the pest species.

Many examples could be cited regarding the effects of blanket applications of a broad-spectrum insecticide on agroecosystems. In 1966, scheduled applications of carbaryl were made in the Imperial Valley of California in an effort to eradicate a recent invader, the pink bollworm, Pectinophora gossypiella (Saunders). Smith {1970} summarized this effort as follows: "Severe and destructive outbreaks of the cotton leafperforator and to a lesser extent spider mites occurred in the Imperial Valley of California as a result of a heavy chemical treatment schedule for cotton to eradicate pink bollworm. Natural enemies of many pests were eliminated and later, when the cotton treatments were stopped, the most severe outbreak ever recorded in this area of cotton loopers and beet armyworms developed on a variety of fall-planted crops." Although eradication was not attempted when the pink bollworm completed its spread across Arizona in 1965, scheduled applications of insecticide to control this pest since 1966 have also produced severe cotton leaf-perforator problems {Watson and Johnson, 1972}.

VALUE OF SELECTIVE VS. BROAD-SPECTRUM INSECTICIDES

The adverse consequences of using broad-spectrum insecticides without regard to all components of the environment have been well documented. Alternative methods have been sought to minimize these adverse effects, including increased emphasis on selective insecticides, particularly the use of insecticides to gain selective action. The phenomenon of selectivity must be utilized not only from the standpoint of protecting beneficial species but also with regard to maximum effectiveness against the pest species. Methyl parathion, applied at the rate of 1.0 lb per acre, is quite effective against the bollworm and tobacco budworm in Arizona while ethyl parathion is relatively ineffective. Unfortunately, a popular formulation consisting of 6 lbs ethyl and 3 lbs methyl parathion per gal is marketed in Arizona. Many applications of this formulation, at the rate of one pint or 1.125 lbs total toxicant per acre, have been made to cotton for bollworm or budworm control. Results have been disastrous since the amount of methyl parathion, 0.375 lb per acre, was insufficient to provide control while the total dosage of both eliminated beneficial species.

Achieving selective action with insecticides has not been easy. In most cases it has required a more intimate knowledge of not only the key pest(s) but also of the secondary pests and beneficial species as well. Selectivity has been achieved by a number of means, but in general, it has been obtained by one of the following practices: (1) use of truly selective materials which are inherently more toxic to some species than to others; (2) timing of applications to suppress pest species while having minimum detrimental effects on natural enemies; and (3) reduced dosages to give adequate control of the target pest while sparing relatively large numbers of natural enemies of the target pest and/or of other potential pests. Pickett {1959} reported that "in apple orchards in Nova Scotia, general use of a program of selective sprays for several years has been followed by a reduction in the amount of damage caused by important insect and mite pests." By replacing DDT with ryania he not only obtained good control of the codling moth, *Laspeyresia pomonella* (L.), but also preserved large numbers of beneficial insects and mites in six major groups. Progress was also made in his studies by reducing the dosage of some of the more toxic chemicals to a point where pest species were controlled and beneficial species harmed but little. Thomas et al. {1959} showed that a glyodin-ryania schedule on apples provided acceptable codling moth control plus satisfactory control of phytophagous mites. Their observations showed that a number of predaceous arthropods known to reduce mite populations appeared within one season following the discontinuance of DDT sprays.

Hoyt {1969} in Washington and Westigard {1971} in Oregon have made significant advances in developing integrated control programs on fruit tree pests. Hoyt's studies on apples showed a wide selection of pesticides that would allow the predaceous mite, *Typhlodromus occidentalis* Nesbitt, to survive in moderate numbers. Even with some of the highly toxic materials, he found that predator mortality could be reduced by using selective spray techniques or in some cases selective timing of sprays. Standard spray programs to control other pests, particularly the codling moth, ultimately resulted in severe spider mite problems.

In his standard spray program, acaricides that usually controlled Tetranychus mcdanieli McGregor, during May, June, and July, but later permitted populations to develop to relatively high levels immediately prior to or during harvest, became ineffective because of mite resistance to the acaricides. It was found that under the standard spray program mite control measures were needed more frequently, at higher dosages, and later in the season. Hoyt {1969} stated, "Thus, the primary function of the acaricide has been to delay rather than control the development of mite populations." High populations at harvest were found to reduce fruit size, produce poor fruit color, and possibly to affect fruit maturity. He found that in the integrated control program, T. mcdanieli was practically nonexistent, or reached its peak (usually at lower levels) and declined well before harvest.

Westigard {1971} encountered similar problems in developing an integrated control program on pears. Among the many problems which had to be considered was the economic injury produced by low population levels of both the pear rust mite, Epitremerus pyri (Nalepa) {Westigard, 1969}, and the two-spotted spider mite, Tetranychus urticae Koch {Madsen and Barnes, 1959}. These low economic injury levels tended to discourage pear growers from allowing the time necessary for the typical lag period between buildup of the mites and of their predators. Another problem was the intense concurrent summer spray programs for control of the pear psylla, Psylla pyricola Foerster, and the codling moth. One of these pest problems was solved by 1967 when an area-wide dormant spray program for pear psylla control replaced the multiple summer sprays. The remaining problem was the commercial practice of controlling codling moths with multiple applications of azinphosmethyl at rates of 1.5 to 2.0 lbs per acre. Insecticidal control of the codling moth was essential to prevent economic losses, however, the dosages used eliminated predatory species. Westigard demonstrated these points in an orchard that had received standard commercial treatment until 1964, at which time all summer sprays were omitted. Two-spotted spider mite population densities peaked at 17, 2.5, 2.5, and 0.18 mites per leaf, respectively, for the succeeding 4 years. This showed that with the cessation of pesticide applications there ensued a trend toward lower average population densities of *T. urticae* and corresponding increases in predatory mite levels. During this period of time, however, the codling moth caused fruit damage averaging about 40 percent, demonstrating the need for chemical control. Subsequent studies with lower rates of azinphosmethyl showed that the reduced rate of 0.5 lb per acre gave adequate codling moth control but still eliminated predatory mites. Still lower rates of 0.125 and 0.25 lb improved survival of *T. occidentalis* but gave inadequate codling moth control. In 1966, he found that azinphosmethyl at 0.125 lb per acre in combination with oil provided the best overall control by combining the effect of oil as acaricide yet with good predator mite survival. Although codling moth control was not acceptable, he felt it did warrant further testing in commercial orchards where moth levels would be considerably lower than encountered in the test orchard. Under those conditions the reduced rates of azinphosmethyl were adequate for control of the codling moth and were selective to the point of allowing increased densities of *T. occidentalis*. During the first year of his study phytophagous mites did increase to unacceptable levels but thereafter, a more favorable predator-prey ratio developed which prevented such increases. This demonstrated the conversion phase of changing from the standard to the integrated program, in which the phytophagous mites exceeded the economic threshold the first year. This again points to the complexity of agroecosystems and that changes from complete reliance upon chemicals cannot be made instantaneously.

METHODS OF IMPROVING SELECTIVITY

Most of the foregoing discussion has indicated that many insecticides do exhibit differential toxicities to various insects, particularly among insect or arthropod groups. From a practical standpoint, however, little of this inherent selectivity is realized when pesticide dosage levels considered necessary to control major pests are applied on a field basis, often with repeated applications. This was indicated in the research of Pickett {1959}, Hoyt {1969}, and Westigard {1971}, where much of the selective action of pesticides was achieved through changes in timing, and numbers and dosages of pesticide applications, accompanied by reevaluations of economic thresholds and influences of predators on their prey.

With the current difficulties in getting registrations for new pesticides it would take several years before new, truly selective pesticides could be available for grower use, assuming they were in existence today! It appears, therefore, that the greatest opportunity of utilizing selective insecticides, at least in the near future, lies with agricultural entomologists and ecologists, who can devise ways and means of using the insecticides currently available in a manner more compatible within the total environment, including agricultural and all other man-centered ecosystems, than has been done in the past.

A number of investigators (such as Pickett, 1959; Hoyt, 1969; and Westigard, 1971) have demonstrated the possibilities of gaining selective advantage in pesticide use. Other approaches to more effective uses of pesticides are beginning to appear in the literature. Efforts are being made to apply greater proportions of the insecticides needed directly on the target crop and less on the surrounding areas. Significant advances in this area alone could result in reduced dosages and still control target pests. Ware et al. {1969} showed that, at all distances downwind, aerial application resulted in 4--5 times as much pesticide drift as that from applications with high-clearance ground sprayers, for both morning and evening treatments. They further showed that aerially applied insecticides in Arizona apparently deposit less than 50 percent of the materials on target during the normal insecticide-use growing season {Ware et al., 1970a}. Another study {Ware et al., 1972} indicated that efficiency of on-target deposits of aerial sprays is influenced by time of day at which applications are made. Ware et al. {1970b} also showed that the addition of certain spray thickeners to an insecticide emulsion reduced the drift from aerial applications. The utility of these results from the standpoint of enhancing selectivity within the target area might be minimal. However, from the standpoint of protecting beneficial species or reducing residues in adjacent nontarget areas the benefits might be manifold.

More precision in the placement and timing of pesticide applications offers hope for gaining selective action. Currently, the pink bollworm in Arizona necessitates repeated spray applications to prevent economic losses in the cotton crop. A decade ago, when it became a general problem across Arizona, control recommendations were rather indefinite as to when applications were needed; dosage levels of insecticides were excessive, e.g., azinphosmethyl was used at 1.0 lb per acre. In recent years, with more precision in timing of applications, based on better estimates of economic population and damage thresholds, and research showing that 0.5 lb per acre of azinphosmethyl gives control comparable to the previously used 1.0 lb rate, maximum protection is now possible with a considerable reduction in the amount of insecticide required {Watson and Fullerton, 1969}. An experiment was conducted using insecticides for pink bollworm control in a much more selective way. This study involved the behavioral characteristics of both the pink bollworm and a sporadic pest complex including the bollworm, *Heliothis zea* (Boddie), and the tobacco budworm, *H. virescens* Fab.

The pink bollworm will oviposit on bolls in preference to squares {Brazzel and Martin, 1957} while the bollworm complex prefers the tender terminal growth on which to oviposit. Young larvae hatching from bollworm eggs will feed for several days on squares in the terminal area before moving down the plant to attack bolls. Studies on population growth of the pink bollworm indicate that, in general, economic infestation levels are reached during the second boll generation, which is usually the latter part of July {Slosser and Watson, 1972}. This corresponds quite closely to the time when bollworm and tobacco budworm oviposition begins. Thus, scheduled applications of insecticides are needed to control the pink bollworm at the time when predatory action against potentially damaging *Heliothis* populations is most needed. In many cases, *Heliothis* outbreaks occur following the initiation of pink bollworm treatments.

The experiment involving selective placement of the broad-spectrum insecticide, azinphosmethyl, demonstrated the possibility of controlling pink bollworms while sparing predators in the terminal portion of the cotton plant. This involved blocking out the over-the-row spray nozzle and utilizing only the two side nozzles to restrict the spray coverage to the lower two-thirds of the plant, where most of the bolls are located when pesticide applications are needed to control the pink bollworm. *Orius* spp., efficient bollworm egg predators {Whitcomb and Bell, 1964}, were particularly abundant in the terminal area of the plants and these, as well as other predatory species, were maintained in the plant terminals at levels comparable to the untreated checks. Pink bollworm control was not as good as in the three-nozzle/row treatment but it was adequate.

The potential for obtaining selective action with currently available insecticides appears to be great. Only a few of the ways have been mentioned. Other possibilities include insecticide-baited traps in conjunction with an attractant, chemically induced sterility, or even the use of other generally more specific types of insecticides, e.g., biotic insecticides or insect pathogens.

Regardless of the means of achieving selectivity, little progress can be made in obtaining selective pesticidal action unless we have an adequate field force of well-trained entomologists schooled in concepts of pest management rather than "pest control." These persons would have intimate knowledge of crops, pest and beneficial species complexes, economic injury levels, relative toxicities of available pesticides, appropriate formulations, application methods, minimum dosages, etc. It may be argued that we have people with these qualifications already on the job and that their efforts allegedly indicate that as yet we are not achieving the goals of selectivity. But it is simply not so, and the number of qualified persons is still too few to adequately cover the broad areas and the many crops involved in the seasonal interactions between pests and their natural enemies. Pest management specialists, who should soon be available to help alleviate this problem, are now being trained in various entomology departments across the country. There must be a more appreciative demand for such individuals if the supply is to continue to be cultivated. This means selling the grower on the pest management approach to crop protection. It also means that this scientifically and economically sound approach toward meeting and reducing the problems and tensions of the present pesticide crisis must be more widely supported by all levels of the chemical industry, from top management to retail sales forces.

REFERENCES

Bartlett, B. R., 1964. Integration of chemical and biological control. *Biological Control of Insect Pests and Weeds* (P. DeBach, ed), Reinhold Publishing Corp., New York, Chap. 17.

Brazzel, J. R., and D. F. Martin, 1957. Oviposition sites of the pink bollworm on the cotton plant. J. Econ. Entomol. *50*:122-124.

Chant, D. A., 1964. Strategy and tactics of insect control. Can. Entomol. *96*:182-201.

DeBach, Paul, 1946. An insecticidal check method for measuring the efficacy of entomophagous insects. J. Econ. Entomol. *39*:695-697.

DeBach, Paul, 1955. Validity of the insecticidal check method as a measure of the effectiveness of natural enemies of diaspine scale insects. J. Econ. Entomol. *48*:584-588.

Falcon, L. A., R. van den Bosch, C. A. Ferris, L. K. Stronberg, L. K. Etzel, R. E. Stinner, and T. F. Leigh, 1968. A comparison of season-long cotton-pest-control programs in California during 1966. J. Econ. Entomol. *61*(3):633-642.

Gasser, R., 1966. Use of pesticides in selective manners. Proc. FAO Symposium on Integrated Pest Control. *2*:109-113, 1965, Rome, Italy.

Getzin, L. W., 1960. Selective insecticides for vegetable leafminer control and parasite survival. J. Econ. Entomol. *53*(5):872-875.

Hall, Irvin M., and Paul H. Dunn, 1959. The effect of certain insecticides and fungicides on fungi pathogenic to the spotted alfalfa aphid. J. Econ. Entomol. *52*(1):28-29.

Hoyt, S. C., 1969. Integrated chemical control of insects and biological control of mites on apples in Washington. J. Econ. Entomol. *62*(1):74-86.

Lingren, P. D., and R. L. Ridgway, 1967 Toxicity of five insecticides to several insect predators. J. Econ. Entomol. *60*(6):1639-1641.

Lord, F. T., 1947. The influence of spray programs on the fauna of apple orchards in Nova Scotia: II. Oystershell scale. Can. Entomol. *79*:196-209.

Lord, F. T., 1949. The influence of spray programs on the fauna of apple orchards in Nova Scotia: III. Mites and their predators. Can. Entomol. *81*(8 and 9):202-230.

Lyle, Clay, 1947. Achievements and possibilities in pest eradication. J. Econ. Entomol. *40*(1):1-8.

MacPhee, A. W., and K. H. Sanford, 1961. The influence of spray programs on the fauna of apple orchards in Nova Scotia: XII. Second supplement to VII. Effects on beneficial arthropods. Can. Entomol. *93*(8):671-673.

Madsen, H. F., and M. M. Barnes, 1959. Pests of pear in California. Calif. Agr. Exp. Sta. Ext. Serv. Circ. 478, p. 40.

Metcalf, C. L., W. P. Flint, and R. L. Metcalf, 1962. *Destructive and Useful Insects*. McGraw-Hill Book Co., New York, p. 62.

Metcalf, R. L., 1966. Requirements for insecticides of the future. Proc. FAO Symposium on Integrated Pest Control. *2*:115-133, Oct. 11-15, 1965, Rome, Italy.

Newsom, L. D., 1966. Essential role of chemicals in crop protection. Proc. FAO Symposium on Integrated Pest Control. *2*:95-108, Oct. 11-15, 1965, Rome, Italy.

Pickett, A. D., 1959. Utilization of native parasites and predators. J. Econ. Entomol. *52*(6):1103-1105.

Reynolds, H. T., V. M. Stern, T. R. Fukuto, and G. D. Peterson, Jr., 1959. Potential use of Dylox and other insecticides in a control program for field crop pests in California. J. Econ. Entomol. *53*(1):72-78.

Ridgway, R. L., P. D. Lingren, C. B. Cowan, Jr., and J. W. Davis, 1967. Populations of arthropod predators and Heliothis spp. after applications of systemic insecticides on cotton. J. Econ. Entomol. *60*(4):1012-1016.

Ripper, W. E., 1956. Effect of pesticides on balance of arthropod populations. Ann. Rev. Ent. *1*:403-438.

Slosser, J. E., and T. F. Watson, 1972. Population growth of the pink bollworm. Ariz. Exp. Sta. Tech. Bulletin 195, p. 32.

Smith, R. F., 1968. Recent developments in integrated control. Fourth Brit. Insect. and Fung. Conf. *2*:464-471, Nov. 20-23, 1967, Brighton, England.

Smith, R. F., 1970. Pesticides: Their use and limitations in pest management. *Concepts of Pest Management*, (R. L. Rabb and F. E. Guthrie, eds.), North Carolina State Univ. Press, Raleigh, N. C., 1970.

Stern, V. M., 1966. Significance of the economic threshold in integrated pest control. Proc. FAO Symposium on Integrated Pest Control. *2*:41-56, Oct. 11-15, 1965, Rome, Italy.

Stern, V. M. and R. van den Bosch, 1959. Field experiments on the effects of insecticides. Hilgardia *29*(2):103-130.

Stern, V. M., R. F. Smith, R. van den Bosch, and K. S. Hagen, 1959. The integrated control concept. Hilgardia *29*(2): 81-101.

Thomas, H. A., H. B. Specht, and B. F. Driggers, 1959. Arthropod fauna found during the first-season trial of a selective spray schedule in a New Jersey apple orchard. J. Econ. Entomol. *52*(5):819-820.

Van den Bosch, R., and V. M. Stern, 1962. The integration of chemical and biological control of arthropod pests. Ann. Rev. Ent. *7*:367-386.

Ware, G. W., B. J. Estesen, W. P. Cahill, P. D. Gerhardt, and K. R. Frost, 1969. Pesticide drift I. High-clearance vs. aerial application of sprays. J. Econ. Entomol. *62*(4):840-843.

Ware, G. W., W. P. Cahill, P. D. Gerhardt, and J. M. Witt, 1970a. Pesticide drift IV. On-target deposits from aerial application of insecticides. J. Econ. Entomol. *63*(6):1982-1983.

Ware, G. W., B. J. Estesen, W. P. Cahill, P. D. Gerhardt, and K. R. Frost, 1970b. Pesticide drift III. Drift reduction with spray thickeners. J. Econ. Entomol. *63*(4):1314-1316.

Ware, G. W., B. J. Estesen, W. P. Cahill, and K. R. Frost, 1972. Pesticide drift VI. Target and drift deposits vs. time of applications. J. Econ. Entomol. *65*(4):1170-1172.

Watson, T. F., and D. G. Fullerton, 1969. Timing of insecticidal applications for control of the pink bollworm. J. Econ. Entomol. *62*(3):682-685.

Watson, T. F., and P. H. Johnson, 1972. Life cycle of the cotton leaf-perforator. Prog. Agric. Ariz. *24*(4):12-13.

Watt, K. E. F., 1963. Dynamic Programming, "Look ahead programming," and the strategy of insect pest control. Canad. Ent. *95*:525-536.

Westigard, P. H., 1969. Timing and evaluation of pesticides for control of the pear rust mite. J. Econ. Entomol. *62*:1158-1161.

Westigard, P. H., 1971. Integrated control of spider mites on pear. J. Econ. Entomol. *64*(2):496-501.

Whitcomb, W. H., and K. Bell, 1964. Predaceous insects, spiders, and mites of Arkansas cotton fields. Ark. Agr. Exp. Sta. Bulletin 690. p. 84.

Chapter 5

STRATEGIES IN THE DESIGN OF SELECTIVE INSECT TOXICANTS

R. M. Hollingworth
Department of Entomology
Purdue University
West Lafayette, Indiana

THE PROBLEM AND SOME REALISTIC GOALS

Recently, economists, with many others, have turned their attention to the costs of environmental pollution and related modern maladies. As a result, the term "externalities" is becoming familiar. Externalities may be defined as those losses or rewards inherent in production or consumption for which no adequate compensation is made in the market. The economics of pest control presents a number of interesting examples of externalities {Headley and Kneese, 1969; Carlson and Castle, 1972}, foremost of which are the hidden costs of pesticide usage. In other words, such unwanted side effects as acute or chronic toxicity to wildlife and man, and damaging effects on natural biological control agents, even where recognized, have not been charged directly to the cost of chemical pest control. Until recently, little economic advantage has accrued to any insecticide because it happened to be free of such spillover effects and, consequently, these properties have played a relatively minor role in decisions to seek, develop, market, and apply new compounds. Quite logically under these conditions the ideal control agent has been regarded as being potent, broad-spectrum, cheap, and reasonably persistent {Persing, 1965; Hansberry, 1968; Metcalf, 1972a}. At the same time many materials with interesting and potentially useful patterns of selective toxicity must have been relegated to the archives for economic reasons of cost, limited spectrum of activity, or marginally reduced potency.

The results of these priorities become immediately obvious if we consider the properties of the insecticides most widely used today. In Table 1 are listed the 18 insecticides produced in the United States at 2 million pounds or more in 1971, as listed in a recent compendium {Johnson, 1972}. Alongside are given the acute oral and dermal LD_{50} values for the rat as indices of hazard to mammals, including man, and the Pest Management Index, devised by Metcalf {1972b} to cover toxicity to other nontarget organisms (fish, birds, and the honeybee) and environmental persistence (half-life) as well as mammalian toxicity. This Index has a scale from 3 (most desirable) to 15 (least desirable) for general pest management purposes. Any attempt to classify what is or is not hazardous in these respects will be arbitrary, since hazard varies with the nature of usage, but a reasonable attempt is made in Table 2. The four numerical degrees of toxic hazard are based on provisional levels suggested by the U. S. Environmental Protection Agency in the administration of the new Federal Environmental Pesticide Control Act {Anonymous, 1973}. The verbal designations (column 1) are the author's.

Applying this classification to the insecticides in Table 1 reveals that of these most common compounds fully 39 percent would be classified as highly or relatively hazardous by the oral route and only 28 percent as very safe. By dermal contact

TABLE 1

Toxicological Properties of the Insecticides Produced in Largest Quantity in the United States in 1971

Compound	Acute LD_{50} (mg/kg: rat)		Pest management index[c]
	Oral[a]	Dermal[b]	
Aldrin	55	98	13.0
Azinphos-methyl	10-18	220	10.0
Azodrin	21	112	----
Bux	1050	400[d]	----
Carbaryl	540	>4000	7.0
Carbofuran	8-14	885[d]	12.0
Chlordane	570	530	7.3
Dasanit	2-10	4.1	----
DDT	113	2510	10.7
Diazinon	350	455	9.7
Disulfoton	12.5	6.0	11.3
Dursban	135	202	9.7
Heptachlor	130	250	12.7
Malathion	1375	>4444	5.3
Methoxychlor	6000	>6000	5.3
Methyl parathion	9-42	67	9.7
Parathion	6-15	6.8	11.0
Toxaphene	60	780	10.0

[a] {Johnson, 1972}.
[b] {Gaines, 1969}.
[c] {Metcalf, 1972b}. Rating scale: 3=most desirable; 15=least desirable.
[d] =rabbit, not rat.

TABLE 2

Classification of the Hazard of the Insecticides Produced in the Largest Quantity in the United States in 1971

	Oral LD_{50} (rat) (mg/kg)		Dermal LD_{50} (rat) (mg/kg)		Pest management index	
	Range	Percent	Range	Percent	Range	Percent
Very hazardous	<10	17	<40	17	>12	13
Relatively hazardous	10–50	22	40–200	17	10–12	27
Relatively safe	50–500	33	200–2000	44	8–10	33
Very safe	>500	28	>2000	22	<8	27

again 34 percent are hazardous and only some 22 percent are classified as very safe. At least 40 percent have properties which may make them poorly compatible with pest management practices. Unfortunately, but significantly, data are generally lacking on the toxicity of these compounds to beneficial insects. However, several have been critized on this account also, which is in keeping with their broad-spectrum properties.

One could argue that these materials are older ones, marketed before the necessity for avoiding toxicity to nontarget organisms was fully appreciated, and soon to be replaced by more sophisticated tools. Unfortunately, this view may be only partly correct. For example, if we group the 94 insecticides listed by Johnson {1972} by the date of their U. S. patent as shown in Table 3 and again consider their oral LD_{50} values to the rat, there is no discernible trend towards development of safer compounds over the last 20 to 25 years. A fairly constant 30 percent are hazardous in each period. Furthermore, since a higher percentage (40 percent) of the most successful of these insecticides fall in this class (Table 2), it appears that a hazardous compound has a better chance of being widely used than a safer one.

This does not leave much ground for complacency and it is clear that in practice, despite their undoubted benefits, insecticides, particularly a dangerous few, have led to death and injury to man and to a variety of toxic effects in the environment, just as the data in the previous tables would predict. Although the extent is not fully known, such accidental poisoning remains a serious problem on a world scale and, it seems, is unlikely to change dramatically in the near future.

Another indicator that there are lessons still to be learnt concerning the externalities of pesticide use is the frequent substitution of the parathions for DDT after its removal from general use. Here we have replaced a compound suspected to be environmentally damaging through excessive persis-

Table 3

Long-term Trends in the Level of Acute Toxicity of Insecticides to Mammals

		Date of U. S. Patent		
	Oral LD_{50} (rat) (mg/kg)	Before 1957 Percent	1957-1966 Percent	1967-1971 Percent
Very hazardous	<10	14	10	13
Relatively hazardous	10-50	16	20	19
Relatively safe	50-500	41	43	44
Very safe	>500	29	28	25
Range of toxicities (mg/kg)		1.2 to 8,170	0.9 to >20,000	5 to >34,000
Number of compounds		38	40	16

tence by compounds harmful in terms of acute toxicity to vertebrates and poor compatibility with beneficial insects. Since safer, if more expensive, replacements do exist for many uses, this decision seems shortsighted and is liable to cause further erosion of public confidence in pesticides as well as more repressive regulation.

Fortunately, it is clear that over the last few years the need for improved selectivity has at last been acknowledged and that, like limited environmental persistence, reasonable selectivity is a basic aim in current development programs. However, the lead time from research bench to field is long and getting longer, and change will come slowly.

If there is a real desire to discover and develop compounds with improved selectivity, two questions must be answered: What type of selectivity is it realistic to aim for and what strategies will lead us to it?

It is not uncommon to read of the necessity for insecticides active against a single pest species. This in most cases is quite unrealistic. Very few pests alone are economically important enough to justify the immense costs of new compound development, particularly since there is the very real risk that the total market is lost if control conditions change (e.g., the onset of resistance in the target species, development of alternative controls). G. K. Kohn clearly describes an example of this dilemma in Chap. 6. Additionally, most crops are attacked by a complex of pests, although one or two may be key ones, and the prospect of applying different species-specific agents against each one is not attractive economically or practically. Finally, and decisively, such monotoxic compounds are very rare {Unterstenhofer, 1970} and are unlikely to be found in general screening programs using a limited number of test species.

A more realistic target, recognizing the economic imperatives and limits to our technical abilities, would be to seek biodegradable compounds active at the phylar or even class levels, i.e., compounds with toxicity limited to arthropods or insects. Barring unforeseen secondary effects, this would provide materials of high safety for man and other vertebrates, plants, and microorganisms. This compromise leaves a substantial group of beneficial or neutral arthropods at risk. However, past experience suggests that if such novel groups of selective toxicants can be found, individual compounds will have differing ranges of specificity. Those with appropriate properties will then be available to fit a given pest management situation or can be used in a way to enhance selective effects. Although this overall goal of devising toxicants of limited selectivity may seem modest enough, a historical perspective suggests that in the past it has not proved easy to accomplish.

Over 20 years ago Ripper et al. {1951} outlined two separate roads to achieving selective action in insecticides. This familiar division remains appropriate:

1. Ecological selectivity in which an intrinsically nonselective agent is applied in such a circumscribed way that the exposure of nontarget organisms is minimized. Examples include specificity of time or place of application, or of formulation (e.g., combination with attractants or baits, microencapsulation), or special chemical properties (e.g., systemic action). This subject is discussed by T. F. Watson in Chap. 4 and it is only necessary here to emphasize the importance and fruitfulness of this approach to selectivity, particularly in view of the certainty of the continued development and use of insecticides with unsatisfactory intrinsic selectivity. Understanding of the ecological basis of insect control should increase rapidly in the future and will present increasing opportunities for this type of selectivity.

2. Physiological selectivity in which nontarget organisms are able to tolerate exposure to the agent while target organisms

succumb. This distinction depends on innate physiological or biochemical differences between the two organisms and will be the major topic addressed here. At this point, we again have two broad and critical choices concerning the most efficient division of resources in the search for new compounds with physiological selectivity:

1. Should we seek completely novel structures based on new modes of action? In this case the role of design is broad-scale and based logically on the known differences between vital sites in the metabolism of target and nontarget groups, or
2. Should we seek to modify existing general groups of insecticides for improved selectivity? In this case the role of design is fine-scale involving molecular manipulations of current compounds, often of a rather minor nature.

There are obviously many facets to this choice. Some major factors favoring the first option are the promise of very high selectivity if the metabolic target is chosen judiciously, better patentability of novel groups, the potential in some cases for rapid onset of resistance to new analogues of currently used structures, together with the discouraging fact that highly selective alternatives to current compounds are already known in many instances, but for various reasons have not been successful in replacing their more hazardous analogs. Finally, there may be the feeling that the potential of older classes of insecticides is close to being worked out. On the other side of the balance favoring the second option are the proven potency and economic viability of known families of insecticides which is difficult to duplicate along with the substantial background of knowledge about their toxicity, metabolism, structure-activity relationships, terminal residues, and environmental compatibility. The low investment requirement in production facilities for similar compounds is a consideration. Also, perhaps, there is a reluctance to commit effort to the speculative, long-term exploration and basic research needed for entirely new materials in the current climate of uncertainty and rising costs of pesticide development. In this light it is reasonable to conclude that both approaches have merit and both are considered below. In general, fine-scale design is more promising in the short term and broad-scale design in the longer term.

Compounds with useful selectivity might be obtained by random screening, partial design, or complete design, depending on the degree to which knowledge of the toxicological systems being attacked is used for predictive purposes. It must be admitted that most of our present selective insecticides have arisen largely by the first route, and that for most purposes complete design of compounds is still beyond our abilities, although this is improving. The major contribution of design lies between these extremes as an aid to directing and improving the efficiency of screening programs towards a defined goal.

STRATEGIES IN DISCOVERING INTRINSICALLY SELECTIVE INSECT TOXICANTS

Many processes, physical and biochemical, govern the dynamics of toxicity of an insecticide and its variation between species which leads to selective toxicity. These include penetration of external and internal barriers, distribution within the body, sites of loss including metabolism, excretion, and interference with the operations of some vital receptor leading to a physiological lesion. These factors have been reviewed thoroughly {O'Brien, 1967; Winteringham, 1969; Metcalf, 1972c} and need little general elaboration. The clear conclusion emerges that of these events two are of overwhelming importance in governing selective toxicity, namely metabolism and action at the target receptor, with penetration barriers also playing a role in selectivity on occasion. In fine-scale modification of existing compounds an understanding of these processes and the likely effect of molecular modifications is a prerequisite. However, in the broad-scale design of new types of toxicants discussed here first, we are concerned initially only with identifying a biological receptor system and attempting to produce chemicals to influence it adversely.

Again there exists a choice of two general classes of effect depending on the type of receptor system influenced: Behavioral toxicants and metabolic toxicants. There is no absolute dividing line between behavioral and metabolic toxicants. By metabolic toxicants are meant those compounds which attack interioreceptors causing a biochemical lesion which then fatally disrupts internal physiology. Behavioral toxicants are compounds that act on exterioreceptors causing a "behavioral lesion" in such a way as to reduce the ability of the insect to survive, adapt, or reproduce.

BROAD-SCALE DESIGN: BEHAVIORAL TOXICANTS

Many possibilities for inducing behavioral toxicity, varying in practicality, have been suggested, the most promising now being techniques employing insect pheromones. This area has been reviewed very fully {Shorey et al., 1968; Beroza, 1970; Wood et al., 1970; Jacobson, 1972; Shorey, 1973} and will not be discussed here. In theory such ideas are very promising and could lead us closer to the ideal of monoselectivity (toxicity against a single species) than any other approach except sterile male techniques. However, not unexpectedly, considerable problems have been met under field conditions in using pheromones either as mass trapping agents or to confuse orientation of the two sexes for mating, and the extent of the future usefulness in these roles is not yet well defined {Shorey, 1972; Marx, 1973}. Frequently there may be a need for a precision in timing, placement, or concentration in using a pheromone, which makes it hard to achieve the requisite high degree of behavioral control over a wide area, something which is not required with many conventional pesticides.

In addition to mating behavior, behaviors essential for oviposition, sociality, dispersal, aggregation, and feeding offer attractive targets. As an illustration of this general approach and its manifold challenges one example, with unfulfilled potential, is worth further exploration here.

Much of insect behavior is comparatively stereotyped, reflecting a nervous system of limited size and complexity. The choice of a suitable food plant is one such behavioral pattern. Acceptability of a host plant involves a complex balance of positive and negative stimuli, many of which are chemical, and is based to a considerable extent on the presence or absence of secondary plant substances {Fraenkel, 1969}. Thorsteinson {1960} and Jermy {1966; 1971} among others have stressed the importance of the absence of feeding deterrents rather than the presence of specific attractants as a major factor influencing host acceptability. This is reasonable if we accept that such deterrents have very commonly evolved in plants as a defense against predators {Dethier, 1970}.

This effect plays an important role in the development of insect resistance in plant breeding by producing strains which are unacceptable to the insect (nonpreference). A second type of resistance, based on the presence of metabolic toxicants in the plant (antibiosis), already has a manmade analog through the use of systemic insecticides which render susceptible strains toxic. There is a distinct possibility that we can also mimic the nonpreference mechanism by the development of synthetic antifeeding compounds.

Antifeeding compounds (feeding deterrents, in the terminology of Dethier et al. {1960} are chemicals which prevent the initiation or continuance of feeding on an otherwise suitable host. They are not necessarily repellants or toxicants. Typically an insect feeding on a treated substrate will make an exploratory probing of the surface, find it unacceptable, repeat this abortive process a number of times at different sites, and then either cease feeding or leave the plant to try elsewhere. This response is so inflexible that the insect may starve to death while sitting on an unrecognized banquet. Antifeeding agents already have some practical application since many mothproofing agents act in part in this way {Wright, 1963}.

Some examples of known antifeeding compounds are shown in Fig. 1. Obviously a wide range of structures are capable of acting in this manner. The first two are naturally occurring agents isolated from insect-resistant plants. Beck and his co-workers {Smissman et al., 1957; Beck, 1960} identified 6-methoxybenzoxazolinone (I) (MBOA) as an antifeeding principle for the first brood of the European corn borer in resistant strains of corn. It is present as the glucoside of a precursor, 2,4-dihydroxy-7-methoxy-1,4-(benzoxazin-3-one)(DIMBOA), which is released enzymatically on injury to the plant by feeding. Slow conversion of DIMBOA to MBOA then occurs. In fact Klun et al. {1967} have claimed that the major antifeeding effect in the plant is due to DIMBOA and possibly its glucoside rather than to MBOA.

More recently juglone (II) has been extensively studied by Norris and his associates {Gilbert et al., 1967; Norris et al., 1970} as an antifeeding compound for Scolytid bark beetles. It has potential, as yet unrealized, in treating elms to prevent attack by the beetle vector of Dutch elm disease. Since it is also effective against the American cockroach {Norris et al., 1971} its spectrum of action may be quite broad.

Other examples of such natural compounds are easily found and very many must await discovery. Thus coumarins and their precursors act as antifeedants for blister beetles in sweet

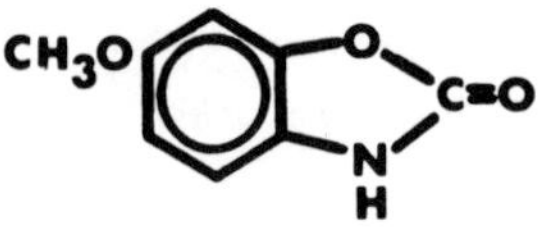

6-Methoxybenzoxazolinone (I)

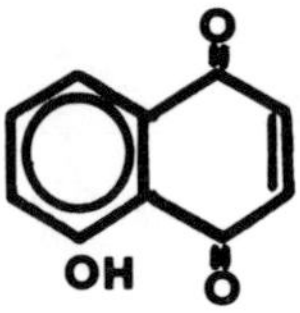

Juglone
5-Hydroxy-1,4-naphthoquinone (II)

American Cyanamid AC-24,055 (III)
4'-(3,3-Dimethyl-1-triazeno)
acetanilide

X=OH Du-Ter (IV)
X=OOCCH3 Brestan (V)

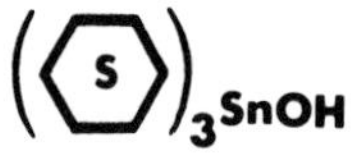

Plictran (VI)

$(CH_3)_3Sn-Sn(CH_3)_3$ Pennwalt TD-3052 (VII)
Hexamethylditin

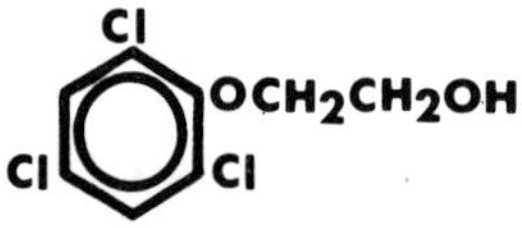

2,4,6-Trichlorophenoxyethanol (VIII)

FIG. 1. Natural and synthetic antifeeding compounds.

clover {Gorz et al., 1972}. Two complex triterpenoids, meliantril {Lavie et al., 1967} and the more potent azadirachtin {Butterworth and Morgan, 1971}, present in the fruit, leaves, and seeds of Meliaceae (e.g., Indian neem tree, chinaberry tree) are effective as locust antifeedants. Earlier studies of the "neem" principle are reviewed by Ascher {1970}. Azadirachtin is particularly impressive since it gave 100 percent antifeeding effect against desert locust nymphs at a concentration of 1 ng per cm^2 of substrate. Its action is claimed to be limited to the Acrididae and it is ineffective against termites, but recent studies have shown it to be active at higher concentrations against some but not all lepidopterous larvae {Ruscoe, 1972}. A partial structure has been described {Butterworth et al., 1972}. A variety of other natural antifeeding agents have been investigated, in some cases with chemical identification of the active principle {National Academy of Sciences, 1972; Munakata, 1970}.

The best known synthetic feeding deterrent is American Cyanamid AC-24005 (III). The properties and field testing of this compound have been presented in detail by Wright {1963; 1967}. In the laboratory it proved to be highly active in deterring attack by many surface feeders but was less effective against those insects that feed only on the deeper tissues, presumably contacting the surface deposit only fleetingly. Success in the field was encouraging in view of the pioneering position of this compound and established that the antifeeding concept is effective and acceptable under some practical conditions. However, for several reasons discussed by Wright, the compound has never been placed on the market. Basically the major drawbacks are the lack of coverage of new growth occurring after the antifeedant is applied, its limitation to surface feeding pests and the marginal cost-effectiveness of the compound. Also, triazenes of this general type have been shown to cause tumors of the nervous system and to be teratogenic in mammals {Preussmann et al., 1969}. The practicability of this antifeedant for field use is still reexamined from time to time {Loschiavo, 1969; Muniappan, 1972}.

A second group of compounds, used in agriculture chiefly as fungicides and acaricides, but with marked antifeeding properties are the organotins, e.g., the fentins Du-Ter (IV) and Brestan (V), Plictran (VI), and Pennwalt TD-3052 (VII). Their properties have been investigated and reviewed by Ascher {1970} (see also Wright {1967} and Jumar {1973}). Briefly, a number of organotins are effective feeding deterrents against a range of surface feeders, in some cases under field conditions. But this has not been sufficient to encourage their commercialization on this basis, though it is conceivable that antifeeding may contribute to their overall effectiveness in some field situations.

The general conclusion from these studies is that the antifeedant approach has much to recommend it. Not least are its high degree of selectivity and compatibility with pest management principles, since natural control agents are not killed directly and hosts are left available to support such parasites and predators. Jermy {1971} has presented the argument that host range is such a strongly fixed character in insects that resistance to antifeedants is unlikely to occur readily, but this remains to be demonstrated. Most antifeedants tested show an encouraging broad range of specificity against phytophagous species and in general seem to be of limited acute toxicity to vertebrates. It has been recognized that, of their major disadvantages, the two most telling, i.e., the need for perfect coverage including new growth after application and the lack of effect on deep-feeding insects, e.g., aphids, could potentially be solved by finding systemic feeding deterrents.

An encouraging factor is that several groups of feeding deterrents contain members with marked systemic action. These include hexamethylditin (VII) {Ascher and Moscowitz, 1969}, the carbamate insecticide Baygon {Matteson et al., 1963} effective at 30 ppm against the boll weevil, and the natural antifeedant azadirachtin, which is active when applied to the soil in 1--10 ppm solution {Gill and Lewis, 1971}. The formamidine insecticide-acaracide chlordimeform has systemic properties and a significant part of its effect in the field may be due to its antifeeding-repellant actions {Doane and Dunbar, 1973}. A particularly interesting series of polyhalophenoxyethanols and acetic acids (e.g., compound VIII, Fig. 1) have been synthesized and field-tested by Jermy and his colleagues {Jermy and Matolcsy, 1967; Matolcsy et al., 1968}. Although these are close relatives of the herbicides 2,4-D and 2,4,5,-T they are not phytotoxic but have strong systemic antifeeding action against such diverse groups as beetles, leafhoppers, and mites. Considerable species specificity was found within the individual compounds of this series. According to Jermy {1971} field tests with these materials have not been conclusive, but the future success of the antifeeding concept probably depends on further developments of this kind.

A crucial element in any attempts to find feeding deterrents by more rational means is an understanding of their mechanism of action. In a few cases we are starting to gain this knowledge and the necessary techniques for studying such processes as ligand-receptor interactions and the electrophysiological background of receptor coding are developing. Norris and his associates have made a comprehensive and fascinating investigation of the antifeeding actions of the naphthoquinone, juglone (II). Studies have been made of structure-activity relationships of analogs {Norris et al., 1970} and binding of naphthoquinones and catechol to a sulfhydryl-containing receptor molecule in the insect antenna in vitro {Norris et al., 1971} and in vivo {Borg and Norris, 1971}. The ultimate outcome of this work may be an understanding of the chemical basis of the receptor-juglone interaction based on reversible oxidation of the sulfhydryl groups in the receptor by the quinone, leading to changes in membrane permeability and potential. Significantly in this redox system the corresponding reduction products, quinols, are feeding stimulants rather than deterrent {Norris, 1970}. Finally, progress has been reported in the isolation of the antennal receptor macromolecule {Ferkovich and Norris, 1972; Rozental and Norris, 1973}.

An entirely different mechanism of deterrence has been suggested by Ascher and Ishaaya {1973} to explain the antifeeding action of the organotin, fentin acetate (V), on larvae of Spodoptera littoralis. It was concluded after extirpation experiments that this compound did not act directly on sensory receptors of the mouth area but it was effective on direct injection into the hemocoel. Antifeeding was considered not to be a consequence of the overall toxic action of organotins. A clear decrease in amylase and protease activity in the gut was observed although the organotins are not direct inhibitors of these digestive enzymes. The basis for the decreased activity and its relationship to antifeeding action remain to be established. The antifeeding compound AC-24,055 (III) was also reported to inhibit protease and amylase activity {Ishaaya et al., 1974}. However, this mode of action does not seem to be consonant with the observations of Wright {1963} who states that AC-24,055 is almost immediate in its effects on insects and is ineffective unless the compound contacts the mouthparts, e.g., no effect is seen after injection. Furthermore, untreated surfaces are fed on at once by larvae that reject treated ones and there appears to be no residual effect on insects exposed to the compound. Thus, there are clear differences in the mode of action of these two compounds. Azadirachtin also has a direct action on gustatory receptors of the desert locust as shown by electrophysiological and extirpation studies {Haskell and Schoonhoven, 1969; Haskell and Mordue, 1969}. A key group of receptors sensitive to azadirachtin were located on the inner surface of the clypeolabrum, but receptors at other sites were also effective so that a high degree of redundancy and plasticity is built into the sensory system.

In conclusion, if there is still much room for debate about any final role for antifeedants in the control of phytophagous pests (or even those that feed on animals), there is enough past success to encourage further vigorous investigation of this concept as a long-range possibility for highly selective control. This role could be either as general agricultural agents or a more specialized one, e.g., in crops where no feeding damage is tolerable or where continuous coverage of the substrate can be guaranteed such as in mothproofing, termite-proofing, or the protection of stored food products. Progress since Wright's review in 1963 has not been rapid but, judging from the open literature, the antifeedant concept has hardly received the concentrated attention devoted to other alternatives such as pheromones and sterilants. Whether it is profitable for industry to energetically pursue the speculative and closely defined goal of finding an antifeedant, especially a systemic one, is doubtful and the past productivity of any such programs has clearly been low, but such a compound should be given the opportunity to show up in general screening programs. Meanwhile, further additions to our rudimentary knowledge of the structure-activity relations and mode of action of feeding deterrents, continuing study of the sensory processes governing feeding, and the isolation and identification of further natural antifeedants will provide an expanding basis for the development of this concept in selective control.

BROAD-SCALE DESIGN: METABOLIC TOXICANTS

Numerous metabolic targets have been discussed in the context of selective toxicity of insecticides, but only a few have been investigated to any depth for their potential in this regard. In fact, there are many cases where sufficient difference exists between analogous receptors present in both insects and nontarget organisms to allow excellent selectivity. An example based on acetylcholinesterase is discussed later. Furthermore, some potential targets of this type are still poorly understood and exploited, e.g., the noncholinergic central and peripheral nervous systems of insects, and these too may offer much scope for selectivity. However, there is an imposing logic to the view that in the long run the most fruitful search for selective toxicants will be directed towards metabolic systems unique or peculiarly important to the target group only.

The biochemistry of arthropods and insects shows many such specializations {Winteringham, 1965; O'Brien, 1967; Twinn, 1972}. Several aspects of the nervous system are promising targets, particularly the neuromuscular apparatus, where the excitatory junctions are probably mediated by glutamate rather than acetylcholine and where inhibitory junctions, perhaps mediated by γ-aminobutyrate, also exist {Pitman, 1971; Gerschenfeld, 1973} in contrast to the vertebrate condition. Other well-defined possibilities include trehalose as the hemolymph sugar and its associated biochemistry, the insect's inability to synthesize sterols and the attendant necessity to modify dietary sterols {Robbins et al., 1971; Bergmann, 1972}, and the α-glycerophosphate shuttle operating in some flight muscle {O'Brien et al., 1965}. Insect hormones also differ greatly in structure and function from those of vertebrates. In the case of the juvenile hormone analogs (JHA) a considerable degree of success has been achieved in exploiting these differences and large-scale field trials are proceeding with these novel, highly selective toxicants. The JHA story is well documented already {Schneiderman et al., 1970; Bowers, 1971; Slama, 1971; Menn and Beroza, 1972; Henrick et al., 1973} and need not be discussed here. They are full of promise for the control of a number of damaging insects but have yet to be proved widely effective in the field. However, the concept of attacking a physiological system limited to certain invertebrates has proved amply correct since so far the JHA have proved virtually innocuous to vertebrates and, in many cases, highly selective between different groups of insects {Henrick et al., 1973; Walker and Bowers, 1973}. Much of the success of the JHA depends on the fact that the natural juvenile hormones are relatively simple and lipophilic. Thus structural analogs (JHA) are toxic as contact insecticides and can be produced economically. Unfortunately, neither of these advantages extends to other insect hormones of known structure (e.g., steroids or polypeptides). This leaves the option of devising selective inhibitors for their synthesis, degradation, or other means to prevent their normal utilization by the insect, but this may be several degrees of magnitude more difficult than making simple structural analogs. Along these lines, Maddrell {1972} has suggested that since a common feature of the action of neurotoxic insecticides is to cause the release of hormones (e.g., the diuretic hormone) which leads to lethal physiological imbalances, these hormones provide a route to selective toxicants. How this can be achieved is not yet clear. In the same vein, various other neurotoxic agents are released in the stressed or poisoned insect {Sternburg, 1963; Pitman, 1971}. However, these compounds have, with few exceptions so far, resisted identification and any development of selective agents based on this observation still lies in the future.

After the JHA, a good candidate for the physiological system most likely to yield selective insecticides is the insect integument. This is a complex, highly organized structure, radically different from that of vertebrates, with functions immediately vital to the survival of the insect. It has at least two major weaknesses:

1. To allow growth it must be shed and resynthesized repeatedly.
2. To prevent rapid desiccation of the terrestrial insect with its large surface-to-volume ratio, it must be effectively waterproofed.

The intricate biochemistry and endocrinology of the integument clearly involves major synthetic activity, storage, and transport mechanisms for carbohydrates to form the polysaccharide chitin, for proteins and orthoquinones derived from tyrosine for sclerotization, and for a range of lipids for waterproofing.

Specialized products, such as the rubbery protein resilin, and antioxidants for cuticular lipids must be produced in some cases. The old cuticle must be digested by specific enzymic secretions, reabsorbed, and the unrecoverable residue shed at ecdysis. During all these events adequate waterproofing must be maintained. The whole process must be initiated and controlled in time and space by precise regulatory mechanisms. At least four hormones are involved, ecdysiotropin and ecdysone which initiate molting, juvenile hormone to control the form of the new cuticle, and bursicon to initiate tanning.

Such a complex of pathways and controls presents a multitude of targets for lethal disruption, and, of course, the JHA and, to a lesser extent, ecdysones are exciting testimonials to the possibilities of success by the hormonal route. However, many other targets are immediately obvious. For instance, can we devise inhibitors for chitinase? If so, the developing insect would be locked in an irremovable armor with no further prospect of growth and maturation. Chitin synthesis is another worthy target. Griseofulvin (IX, Fig. 2) is an antibiotic with uses as an agricultural fungicide. Speculation that its fungicidal action was based on interference with the synthesis of fungal chitin (a hypothesis now in doubt) led Anderson {1966} to study its effects on insects. He found that at less than 20 ppm it caused gross deformation of the cuticle of mosquito larvae in the subsequent instar, including failure of muscles to attach to the cuticle. Chambers and Love {1962} also have reported morphological effects in mites treated with griseofulvin, such as lack of pigmentation and malformation of the abdominal cuticle.

An analogous situation may exist with the polyoxin complex of nucleoside fungicides which bear a structural relationship to UDP-N-acetylglucosamine {Hori et al., 1971}. Polyoxin D has been shown to be a potent competitive inhibitor of fungal chitin

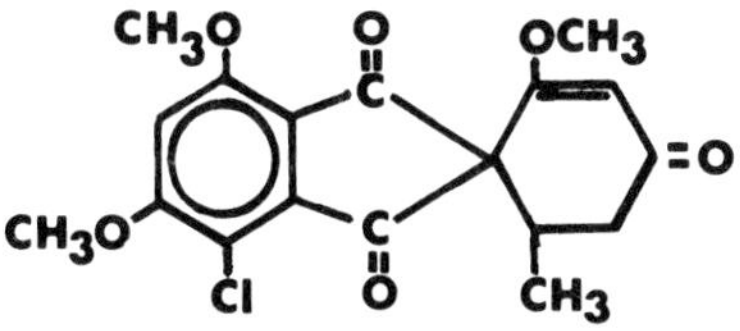

Griseofulvin (IX)

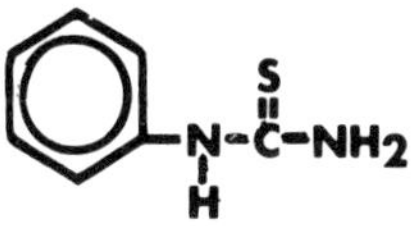

1-Phenyl-2-thiourea (X)

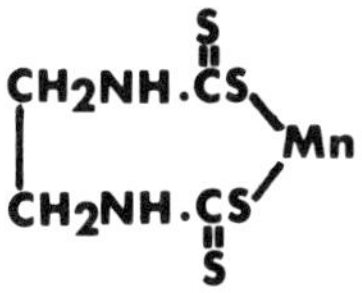

Maneb (XI)

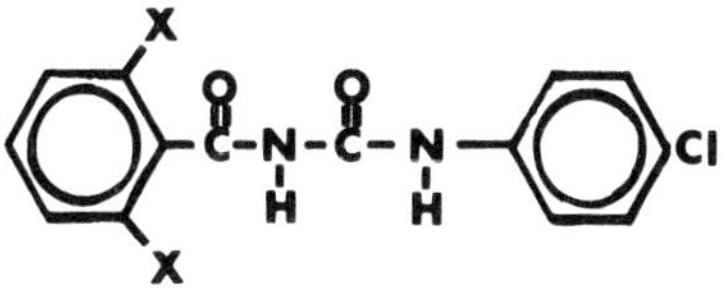

X = Cl TH-6038 (XII)
X = F TH-6040 (XIII)
1-Benzoyl-3-phenylureas

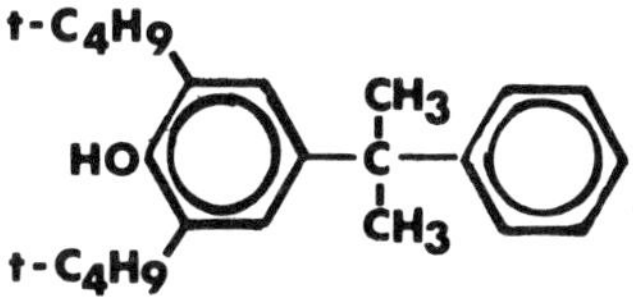

MON-0585 (XIV)

FIG. 2. Some compounds active against the insect cuticle and molting.

synthetase, which explains its inhibitory effect on the synthesis of cell wall chitin {Endo and Misato, 1969}. Polyoxin A has marked insecticidal activity with cuticular involvement e.g., LD_{50} after injection into grasshoppers is 5 μg per insect {A. Morello, personal communication}.

The phenoloxidases of the hemolymph and cuticle necessary for the production of tanning quinones from aromatic amino acids are another interesting prospect. Inhibition of these enzymes should lead to failure to harden and darken the cuticle. There is considerable evidence that this is the case. Thioureas and related compounds (dithiocarbamates, 2-thiouracils) have long been known to be phenoloxidase inhibitors in vitro {Lerner and Fitzpatrick, 1950} presumably by virtue of their ability to complex the copper in these metalloenzymes. Corresponding inhibition of cuticular development and pigmentation has often been seen in treated insects, and phenylthiourea (PTU,X, Fig. 2) has been evaluated as an insecticide, e.g., for control of clothes moths and mosquito and housefly larvae {Ogita, 1958}. Wallis {1961} showed that mosquito larvae lacked melanin and had a much prolonged larval period when exposed to PTU at high concentration (1 mM). McFarlane {1960} observed the action of several thioureas and sodium diethyl dithiocarbamate on cricket eggs and concluded that the toxic action was related to a decreased level of phenoloxidase, which prevented normal tanning of the chorion and subsequent water uptake. Two further studies have related reduced phenoloxidase activity in vivo with toxicity due to cuticular maldevelopment. Edelman and Posnova {1970} studied the effect of feeding 6-methylthiouracil and 1-methyl-2-mercaptoimidazole to a range of insects. No toxic effects were seen until the subsequent molt when there was a failure to shed the old cuticle and, in some species, the new cuticle was poorly pigmented and sclerotized. A 30--60 percent decrease in total phenoloxidase activity was found in the treated insects. Very similar effects on molting and cuticular synthesis were reported by Chmurzynska and Wojtczak {1963} on injecting thiourea into silkworm larvae and again a strong reduction in phenoloxidase activity was considered to be the cause.

Various dithiocarbamate fungicides have been shown to have significant potency in preventing molting in insects, although the mechanism has not been defined, e.g., Ziram (zinc dimethyl dithiocarbamate) at 5--10 ppm strongly delayed pupation of mosquito larvae {Lewallen, 1964} and Maneb (XI, Fig. 2) and Zineb (manganese and zinc ethylenebisdithiocarbamates) had the strange effect of completely preventing molting of greenhouse whiteflies without other untoward effects until the insects became bloated and died {McMullen, 1959}. Maneb in this case had comparable potency to the positive control compound, perthane. The basis of the toxicity of Maneb was investigated further by McMullen {1965} and is clearly complex with a variety of cellular injuries through effects on sulfhydryl enzymes and perhaps metal chelation.

The phenoloxidases represent only one target which can be successfully attacked. A host of other possibilities remain. For example, DOPA decarboxylase converts DOPA to dopamine in the production of tanning quinones. Inhibitors of this enzyme are known e.g., 3-(3,4-dihydroxyphenyl)-2-hydrazino-2-methylpropionic acid completely blocked sclerotization of the puparium of _Sarcophaga bullata_ at 5 μg per pupa with lethal results {Bodnaryk, 1970}. Different genera of Diptera have unique forms for sequestering the large amounts of aromatic amino acids needed for tanning e.g., β-alanyl-tyrosine (Sarcophaga), α -glutamylphenylalanine (Musca) and tyrosine-O-phosphate (Drosophila) {Bodnaryk, 1972}. Specific enzymes for the synthesis and cleavage of these storage forms are present. Clearly targets of this type may present an opportunity for highly specific insecticides if such are desired. A number of investigators have described examples of the precocious onset of tanning of the new cuticle before the insect has completed ecdysis and expansion to its new form. This has the laudable consequence of locking the insect into an inappropriate shape. Actions of this type have been observed with JHA {e.g., Spielman and Skaff, 1967}, with ecdysones {e.g., Bodnaryk, 1971}, and with saturated fatty acids {Quraishi, 1971}. Presumably, disruption of a variety of targets at different stages of the molting cycle could have the same result, e.g., induction of premature synthesis of tanning agents, premature release of bursicon, or interference with cellular permeability for diphenols {Koeppe and Mills, 1972}.

Compounds which interfere with the biochemistry of molting have the limitation of acting only on immature insects. However, an agent which would disrupt waterproofing of the cuticle would not be limited in this respect. A well-known example is the use of inert abrasive or wax-adsorbent dusts, e.g., silica aerogels for control of household and grain insects {Ebeling, 1971}. The hazard of desiccation which faces many insects is clearly shown by the observation of Turpin and Peters {1971} that corn rootworms prefer clay over sandy soil because the latter rapidly abrades the cuticular lipids, leading to death. Knowledge of the biochemistry and control of waterproofing and the nature of repair processes after injury is scanty {Jackson and Baker, 1970}, although there are indications that distinct classes of cuticular hydrocarbons are secreted in adult insects which may be associated with cuticular repair {Jackson et al., 1974}. The continuing production of cuticular lipids appears to be under direct hormonal control {Locke, 1965}. Additional physiological mechanisms may be crucial in controlling water movement through the integument, e.g., Winston and Beament {1969} have cited evidence for an energy-requiring cuticular "water pump" located in the epidermis which lowers the water tension in the cuticle in comparison to the hemolymph.

Examples of interference with waterproofing in insects other than lipid removal by abrasion, sorption, or by surfactants are not plentiful, but Cline {1972} has reported that certain fatty amines have an unusual potency in destroying permeability barriers in the cuticle of mosquito eggs, which can lead to lethal desiccation. Another provocative observation was made by Sun and Johnson {1972}, who found that several neurotoxicants led to rapid water loss in treated insects before any other symptoms of toxicity appeared. They suggested that these compounds may be interfering with neurally controlled mechanisms normally responsible for limiting water loss through the integument.

Although they do indicate that the insect cuticle is a valid site of attack for novel insecticides, many of these examples of interference with cuticular processes may be classed as laboratory curiosities, sometimes of dubious etiology, and initiated by compounds of limited potency. At best they might be the starting points in a long and uncertain road to a commercially useful product. However, there are recent examples of success in finding effective insecticides to interfere with cuticular biochemistry which should encourage further efforts in this area.

A series of 1-benzoyl-3-phenylureas have been described by van Daalen et al. {1972}, orginating from a herbicide synthesis program, which have drastic effects on insect molting. These compounds are not toxic to adult insects nor to immatures by topical application, and their toxicity to vertebrates and, fortunately, to plants is also low. However, after ingesting the compound, immatures from several orders fail to emerge successfully from the exuvia at the subsequent molt and die. This has been attributed to interference with the process of cuticular deposition and particularly with a failure to lay down a normal endocuticle {Mulder and Gijswijt, 1973}. It is claimed {Thomson-Hayward Chemical Co., 1973} that these compounds do not interfere with hormonal regulation of molting, and they are clearly unlike JHA in affecting larval-larval molts. It has recently been suggested that they inhibit chitin synthesis in lepidopterous larvae {Post et al., 1974} and that they stimulate activity of chitinase and cuticular polyphenoloxidase in larval houseflies {Ishaaya and Casida, 1974}. Either, or both, of these actions would lead to a thin, weakened cuticle. Considerable detail of the structure-activity relations of this group has been published recently {Wellinga et al., 1973} and it reveals interesting variations in sensitivity between different species. Two compounds shown in Fig. 2, TH-6038 (DU-19111)(XII) and particularly TH-6040 (XIII) have been chosen for evaluation in the United States.

A second commercially interesting compound with high selectivity and a clear action on insect cuticle is 2,6-di-*t*-butyl-4-(α,α-dimethylbenzyl)phenol (MON-0585) (XIV, Fig. 2) developed as a mosquito larvicide {Sacher, 1971a}. This compound has little activity on insects other than mosquitoes and appears to carry minimal risk for beneficial insects and vertebrates. As with the previous compounds, treated insects show no immediate ill effects. However, when they reach the early stages of pupation the pupae die in an intermediate, unmelanized form. Adults are not affected directly. The timing of the effect is quite different from many known JHA, which block adult emergence rather than pupation {Schaefer and Wilder, 1972}. It has been suggested that MON-0585 may be acting as a phenoloxidase inhibitor {Sacher, 1971b} but its activity in this regard is only modest and the mode of action which remains unknown could occur equally as well at the hormonal level.

In conclusion, there are a variety of vulnerable targets in the biochemistry of insects, many of which are of unique importance to insects or to arthropods in general. Selective toxicants based on such targets are now showing promise of commercial feasibility and there is reason to hope many future successes lie in this direction. It is no hard task to suggest useful targets of this kind in insects and more will become apparent in time. There is every reason to believe that by appropriate, if time-consuming, study we can understand them well enough to make intelligent assessments about how to exploit them. However, there remains one major obstacle which has severely limited success in achieving this goal. Our ability to rationally design agents with requisite properties of selectivity, lipophilicity, potency, and reasonable cost to interfere with known targets has been distressingly limited. It is significant that although the JHA are compounds designed to perform a specific function, both the benzoylureas and MON-0585 were the children of chance and not intellect. Creative bioorganic chemistry is the weak link which must be strengthened if we are to understand the molecular operations of promising receptors and design effective agents for their disruption.

FINE-SCALE DESIGN - THE MODIFICATION OF EXISTING INSECTICIDES

The literature of selectivity is sprinkled with such terms as "tailor-made" insecticides or "precisely-designed" toxicants in reference to agents intended to exactly fit a particular niche in insect control. Intriguing though this idea may be, it is far from realization. The reason is abundantly clear. The nature and interaction of events during poisoning and the environmental influences on the activity of pesticides are usually too complex, too poorly understood and quantitated to allow any highly precise relationship of structure to selectivity to be predicted over a range of compounds and organisms. For instance, there are few helpful generalizations as yet possible about the distribution and properties of the enzymes which metabolize insecticides in vertebrates and insect species {Hollingworth, 1971}. Despite many years of study of acetylcholinesterase (AChE) as a target for organophosphorus and carbamate insecticides, no easy rules have emerged to indicate how one can design an inhibitor selective against the insectan versions of this enzyme, although some helpful leads are available. It is perhaps, then, not surprising that we often do not anticipate either the appearance of selectivity when it does arise or, equally, the failure of selectivity in an apparently favorable situation. The power of structure-activity studies has increased considerably in the past few years, notably by the successes of the Hansch approach, but they have not yet been widely used in studying selectivity. At the same time knowledge of comparative toxicology has accumulated rapidly and there are increasing opportunities to develop improved selectivity on a rational basis. Generally this takes the form of "design by analogy," in which known selective agents act as a pattern on which new derivatives can be based, often by the inclusion of a selectophoric group present in the parent molecule, but sometimes, as with the promising new synthetic pyrethroids {Elliott, 1971}, by molecular modification within a class whose members are almost invariably of low mammalian toxicity.

It was earlier suggested that of the manifold events influencing toxicity, differences in rates of metabolism and in properties of the receptor site between organisms are likely to have the greatest influence on selectivity. Space does not permit a detailed discussion of these factors and their role in selectivity, but one brief example from each of the two areas will serve to illustrate how fine-scale design based on observed differences between insect and vertebrate reponse to known compounds can lead to promising results.

The effect whimsically termed the "magic meta methyl" {O'Brien, 1967} has been discussed on several occasions {Schrader, 1961, 1965; Metcalf, 1972c; Metcalf and Metcalf, 1973}. It refers to the enhancement of selectivity between mammals and insects when a methyl (or chlorine or trifluoromethyl) group is inserted in the meta position of the phenyl ring of some organophosphates. A striking and familiar example is the modification of the inadequately selective methyl parathion (*O*,*O*-dimethyl *O*-*p*-nitrophenyl phosphorothioate) to fenitrothion (*O*,*O*-dimethyl *O*-3-methyl-4-nitrophenyl phosphorothioate) which is excellently selective (Table 4). The reason for this startling enhancement in selectivity by such a modest structural change is a mystery which has been widely investigated {Vardanis and Crawford, 1964; Hollingworth et al., 1967a, 1967b; Douch et al., 1968; Miyamoto, 1969}, incidentally without providing any wholly convincing explanation. Our own studies pointed to differences in the ability of the respective phosphate analogs, methyl paraoxon and fenitroxon, to inhibit mammalian and insect AChE as being one, but by no means the only, factor in the improved selectivity. A simple model based on differences in a hydrophobic binding site adjacent to the esteratic site of AChE was proposed to explain the differential inhibition. The toxicities of methyl parathion and fenitrothion to the mouse and the housefly are shown in Table 4, together with the anticholinesterase activity of their oxons. The ratio of anticholinesterase activities against insect and vertebrate is also shown (Selective Inhibitory Ratio, SIR). The meta methyl group clearly enhances reactivity with the housefly enzyme and decreases it with the two mammalian enzymes. This leads to an improvement in the SIR of 15 to 20-fold. The Selective Toxicity Ratio (STR) is also improved about 20-fold from 19 for methyl parathion to 403 for fenitrothion. This general observation has been confirmed and extended to include avian AChE by Mehrotra and Phokela {1969}. Their data for the locust, sparrow, and rat in Table 4 again show that the introduction of the meta methyl group enhances the SIR in favor of vertebrates, especially in the case of the sparrow.

Recently Eto et al. {1972} examined the properties of p-acetylphenyl phosphate esters as insecticides. O,O-Diethyl O-p-acetylphenyl phosphorothioate was found to be a reasonably effective toxicant. In keeping with the fenitrothion precedent, the effect of inserting a meta methyl group was examined. As shown in Table 4, this again had the effect of favoring attack on the insect AChE as shown by a 25-fold improvement in SIR.

TABLE 4

The Selective Effect on Anticholinesterase Activity and Toxicity of the Presence of a Methyl Group in the Ring of some Substituted-Phenyl Organophosphates

$$(R^1O)_2\overset{X}{\overset{\|}{P}}-O-C_6H_3(R^3)R^2$$

Structural modifications			Anticholinesterase activity[a]					Acute toxicity[b]		
						SIR[c]				STR[d]
R^1	R^2	R^3	A	B	C	A/B	A/C	D	E	E/D
			Fly	Bovine erythrocyte	Mouse brain			Fly	Mouse	
Me[e]	4-NO_2	3-H	2.9	5.2	1.1	0.56	2.6	1.2	23	19
		3-Me	7.6	0.73	0.18	10.4	42	3.1	1250	403
			Locust	Sparrow	Rat				Rat	
Me[f]	4-NO_2	3-H	8	144	15	0.06	0.53	--	19	--
		3-Me	4	6	3	0.67	1.33	--	250	--

			Fly	Human erythrocyte		Fly	Mouse	
Et[g]	4-$COCH_3$	3-H	27	0.45	60	25-50	200	4-8
		3-Me	77	0.05	1540	16	>200	>8
		2-Me	0.12	0.02	6.0	>500	>200	--
Me[g]	4-$COCH_3$	3-H	--	--	--	25-50	>200	>4-8
		3-Me	--	--	--	8.5	1500	176

[a] Activity expressed as $k_i x10^{-5} (M^{-1} min^{-1})$ with X=O.

[b] Toxicity expressed as LD_{50} in mg/kg with X=S.

[c] SIR = Selective Inhibitory Ratio, Insect/Vertebrate.
[d] STR = Selective Toxicity Ratio, Vertebrate/Insect.
[e] {Hollingworth et al., 1967a}.
[f] {Mehrotra and Phokela, 1969}.
[g] {Eto et al., 1972: k_i calculated from I_{50}; LD_{50} based on 50 flies/g}.

Unfortunately the authors do not present complete details of the mammalian toxicity of these compounds but the meta methyl analog is shown as less toxic to rats and of increased toxicity to flies, so that the STR has again moved in the direction indicated by the change in SIR. By comparison with the parathions, one would expect an even greater effect on selectivity with the dimethyl esters in this series (R^1=Me in Table 4). O,O-Dimethyl O-3-methyl-4-acetylphenyl phosphorothioate was the least toxic compound to the mouse (LD_{50}=1500mg/kg) and was more toxic than the diethyl derivative to the fly (LD_{50}=8.5mg/kg) to given an overall STR of 176. This molecular modification has again produced a highly selective compound. Interestingly, moving the methyl group from the 3-position to the 2-position of the ring in these compounds decreases the SIR by 10-fold, primarily by strongly decreasing the reaction with the fly enzyme. Reasonably this also leads to much reduced insecticidal activity. To close this circle of analogy, we see a similar effect with the p-nitrophenyl series {Metcalf and Metcalf, 1973} where the 2-methyl analog is 5- to 10-fold less active as an inhibitor of fly AChE than the unsubstituted methyl paraoxon and correspondingly is markedly inferior as a fly toxicant compared to either methyl parathion or its 3-methyl analog, fenitrothion. It is hard to imagine a more impressive example of the immense potential for the improvement of selectivity by fine-scale design than the effect of these simple manipulations with a humble methyl group.

The second example, this time based on metabolism as the selective force, is concerned with N-methyl carbamate insecticides. These provide examples of some of the most mammalitoxic (e.g., aldicarb, carbofuran) and some of the most selective (e.g., carbaryl, butacarb) insecticides. A significant improvement in general selectivity between insects and mammals in many members of this class was reported by Fraser et al. {1965} through synthesis of the corresponding N-acetyl N-methyl analogs.

This effect is shown in Table 5 for m-isopropylphenyl N-methylcarbamate. A 15- to 30-fold decrease in toxicity to the mouse is obtained while the effect on toxicity to insects ranges from a strong decrease to enhanced activity in the case of the mosquito larvae. Changes in anticholinesterase activity do not explain this effect since N-acylation leads to a considerable and general decrease in anticholinesterase activity. A plausible explanation has been provided by Miskus et al. {1969} who studied the metabolism of the N-acetyl derivative of Zectran in the mouse and spruce budworm. Previous studies {Miskus et al., 1968} had shown that compared to Zectran, the N-acetyl derivative was much less toxic to the mouse, but only slightly less toxic to budworm larvae. In the budworm, deacetylation occurred to produce significant levels of the potent parent, Zectran. Mice, on the other hand, produced little free Zectran but degraded the N-acetyl derivative largely to innocuous phenolic products. Thus, the improved selectivity appears to depend on a more efficient lethal synthesis of Zectran from its N-acetyl derivative in the insect than the mammal.

Following on these studies a range of analogously N-substituted derivatives of familiar carbamates have been prepared (Table 5). These include a series of N-phosphorylated carbamates {Fahmy et al., 1970}, alkyl- and arylsulfenylated carbamates {Schaefer and Wilder, 1970; Black, Chiu, Fahmy, and Fukuto, 1973} and a series of biscarbamoyl sulfides in which two carbamates were linked by a sulfide bridge between the carbamyl nitrogen atoms {Fahmy et al., 1974}. As the data in Table 5 reveal, these varied derivatives show a remarkable similarity in their toxicological properties. Like the N-acetyl analog, in each case mammalian toxicity is significantly decreased while in many cases insecticidal activity is maintained or enhanced. In particular, toxicity to mosquito larvae is improved, perhaps as a result of the enhanced lipophilicity of the derivatives

TABLE 5

Selective Effects on Anticholinesterase Activity and Toxicity by N-derivatization of _m_-isopropylphenyl _N_-methylcarbamate

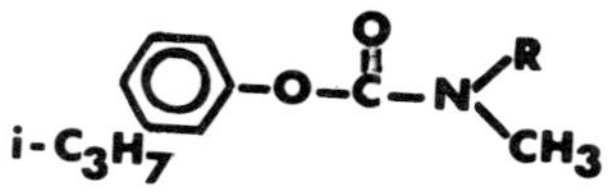

		R=H	$R=CH_3CO-$	$R=(CH_3O)_2P(S)-$	$R=C_6H_5S-$	Biscarbamoyl sulfide
Anticholinesterase activity in vitro	Housefly[a]	6.5[c]	4.4[d]	4.2[c]	--	---
	Honeybee[a]	7.7[e]	4.7[e]	--	--	---
	Bovine erythrocyte[b]	7.5[i]	--	--	1.2[i]	0.27[j]
	Housefly[b]	7.7[i]	--	--	0.37[i]	0.23[j]
Toxicity (LD_{50})	Mouse (mg/kg)	16[c]	250-500[f]	760[c]	150-200[i]	200[j]
	Fly (Sus.)(mg/kg)	41[c]	235[d]	32.5[c]	75[i]	85[j]
	Fly (Res.)(mg/kg)	125[c]	--	33.5[c]	--	---
	Culex larva (ppb)	38[d]	28[d]	--	6[i]	5.6[j]
	Blowfly larva (ppb)	<30[f]	200[f]	--	--	---

Aedes aegypti (ng/adult)	3.6[g]	4.2[g]	--	--	---
Anopheles stephensi (ng/adult)	2.0[g]	2.8[g]	--	--	---
Mexican bean beetle (ppm)	10[h]	>100[h]	--	--	---
Bean aphid (ppm)	30[h]	>1000[h]	--	--	---

[a] pI_{50}(M)
[b] $k_i \times 10^{-5}$ ($M^{-1}min^{-1}$)
[c] {Fahmy et al., 1970}.
[d] {Fahmy et al., 1966}.
[e] {Lewis, 1967}.
[f] {Fraser et al., 1967}.
[g] {Hadaway et al., 1970}.
[h] {Weiden, 1971}.
[i] {Black, Chiu, Fahmy, and Fukuto, 1973}.
[j] {Fahmy et al., 1974}.

{Black, Chiu, Fahmy, and Fukuto, 1973}. In each instance, anticholinesterase activity of the derivatives is diminished compared to the parent compound. Metabolic investigations in the case of an N-arylsulfenyl derivative of carbofuran in the housefly and mouse {Black, Chiu, Fukuto, and Miller, 1973} led to the same conclusion obtained with N-acetyl Zectran since large amounts of free carbofuran were found in the insect but not in the mouse. A similar explanation may be forthcoming in the case of the N-phosphorylated carbamates {Fahmy and Fukuto, 1972}. The fact that metabolic activation is a key to toxicity may also explain why the derivatized carbamates also show, in some cases, favorable toxicity against resistant insects in which insecticide metabolism is enhanced, e.g., houseflies (Table 5), mosquitoes {Schaefer and Wilder, 1970} and Egyptian cotton leafworms {Fahmy and Fukuto, 1970}. From these studies emerges a coherent picture of successful manipulation of toxicity to enhance selectivity, with no indication that the potential of N-derivatization of the carbamates has been exhausted.

CONCLUSION

These examples illustrate that there is still much scope for rational modification of our present families of insecticides and that the potential for devising new types to attack well-chosen target receptors remains largely untapped. The central question must then be, if such compounds are found, will they be used? It is unfortunate that frequently an increase in selectivity is accompanied by a contraction of the spectrum of activity and some decrease in potency against important pests. Since the more selective compounds are also often more expensive to synthesize, they have not, under past economic and regulatory incentives, been very successful in displacing established less selective agents. The materials discussed in the final section are a perfect illustration of this most critical prob-

However, its use in the United States has been very limited although its much more dangerous sibling, methyl parathion, has seen a rapid growth in importance in the last decade. The derivatized carbamates are more recent, but have not yet been used commercially and, judging from the material presented by Dr. Kohn in Chap. 6, may have problems in penetrating a significant part of the market in spite of their outstanding properties, e.g., as mosquito larvicides.

The foremost strategy in devising selective insecticides is therefore not one of sophisticated biochemistry but simply to find ways to put a high value and reward on selectivity and to penalize those dangerous, broad-spectrum agents with the highest external costs. Unfortunately, although through burgeoning pesticide regulation the latter may be accomplished, the same process tends to so increase the costs and uncertainties of pesticide development that selective and narrower-spectrum compounds become unattractive and long-term innovation is discouraged. This is a paradox which urgently needs resolution. Otherwise many promising advances in selectivity will remain laboratory exercises without practical application.

REFERENCES

Anderson, J. F., 1966. J. Econ. Entomol. *59*:1476-1482.

Anonymous, 1973. Pestic. Chem. News. *1*(36):12-16.

Arnold, M. T., L. L. Jackson, and F. E. Regnier, 1973. Abstract of paper presented to Amer. Chem. Soc., Pesticide Chem. Div., Chicago, Illinois.

Ascher, K. R. S., 1970. World Rev. Pest Control. *9*:140-155.

Ascher, K. R. S., and I. Ishaaya, 1973. Pestic. Biochem. Physiol. *3*:326-336.

Ascher, K. R. S., and J. Moscowitz, 1969. Internat. Pest. Contr. *11*:17-20.

Beck, S. D., 1960. Ann. Entomol. Soc. Amer. *53*:206-212.

Bergmann, E. D., 1972. Proc. 2nd Internat. IUPAC Congr. Pestic. Chem. (A.S. Tahori, ed.), *1*:1-12.

Beroza, M., 1970. *Chemicals Controlling Insect Behavior*, Academic Press, New York.

Black, A. L., Y. C. Chiu, M. A. H. Fahmy, and T. R. Fukuto, 1973. J. Agr. Food Chem. *21*:747-751.

Black, A. L., Y. C. Chiu, T. R. Fukuto, and T. A. Miller, 1973. Pestic. Biochem. Physiol. *3*:435-446.

Bodnaryk, R. P., 1970. Comp. Biochem. Physiol. *35*:221-227.

Bodnaryk, R. P., 1971. Gen. Comp. Endocrinol. *16*:363-368.

Bodnaryk, R. P., 1972. Comp. Biochem. Physiol. *43B*:587-592.

Borg, T. K., and D. M. Norris, 1971. Ann. Entomol. Soc. Amer. *64*:544-547.

Bowers, W. S., 1971. *Naturally Occurring Insecticides*, (M. Jacobson and D. G. Crosby, eds.), Marcel Dekker, Inc., New York, pp. 307-332.

Butterworth, J. H., and E. D. Morgan, 1971. J. Ins. Physiol. *17*:969-977.

Butterworth, J. H., E. D. Morgan, and G. R. Percy, 1972. J. Chem. Soc. (Perkin I):2445-2450.

Carlson, G. A., and E. N. Castle, 1972. *Pest Control Strategies for the Future*, National Academy of Sciences, Washington, D. C., pp. 79-99.

Chambers, E. E., and F. Love, 1962. Small Anim. Clin. *2*:91-93.

Chmurzynska, W., and L. Wojtczak, 1963. Biol. Bull. *125*:61-68.

Cline, R. E., 1972. J. Econ. Entomol. *65*:177-181.

Dethier, V. G., 1970. *Chemical Ecology*, (E. Sondheimer and J. B. Simeone, eds.), Academic Press, New York, pp. 83-102.

Dethier, V. G., L. B. Browne, and C. N. Smith, 1960. J. Econ. Entomol. *53*:134-136.

Doane, C. C., and D. Dunbar, 1973. J. Econ. Entomol. *66*:1187-1189.

Douch, P. G. C., G. E. R. Hook, and J. N. Smith, 1968. Austr. J. Pharm. *49*:S70-S71.

Ebeling, W., 1971. Ann. Rev. Entomol. *16*:123-158.

Edelman, N. M., and A. N. Posnova, 1970. Doklady Adak. Nauk SSSR. *190*:1004-1007.

Endo, A., and T. Misato, 1969. Biochem. Biophys. Res. Comm. *37*:718-722.

Eto, M., M. Sakata, and T. Sasayama, 1972. Agr. Biol. Chem. *36*:645-660.

Fahmy, M. A. H., Y. C. Chiu, and T. R. Fukuto, 1974. J. Agr. Food Chem. *22*:59-62.

Fahmy, M. A. H., and T. R. Fukuto, 1970. J. Econ. Entomol. *63*:1783-1786.

Fahmy, M. A. H., and T. R. Fukuto, 1972. Tetrahedron Lett. *1972*:4245-4248.

Fahmy, M. A. H., R. L. Metcalf, T. R. Fukuto, and D. J. Hennessy, 1966. J. Agr. Food Chem. *14*:79-83.

Fahmy, M. A. H., T. R. Fukuto, R. O. Myers, and R. B. March, 1970. J. Agr. Food Chem. *18*:793-796.

Ferkovich, S. M., and D. M. Norris, 1972. Experientia *28*: 978-979.

Fraenkel, G., 1969. Entomol. Exp. Appl. *12*:473-486.

Fraser, J., P. G. Clinch, and R. C. Reay, 1965. J. Sci. Food Agr. *16*:615-618.

Fraser, J., D. Greenwood, I. R. Harrison, and W. H. Wells, 1967. J. Sci. Food Agric. *18*:372-376.

Gaines, T. B., 1969. Tox. Appl. Pharmacol. *14*:515-534.

Gerschenfeld, H. M., 1973. Pharmacol. Rev. *53*:1-119.

Gilbert, B. L., J. E. Baker, and D. M. Norris, 1967. J. Ins. Physiol. *13*:1453-1459.

Gill, J. S., and C. T. Lewis, 1971. Nature (London). *232*: 402-403.

Gorz, H. J., F. A. Haskins, and G. R. Manglitz, 1972. J. Econ. Entomol. *65*:1632-1635.

Hadaway, A. B., F. Barlow, J. E. H. Grose, C. R. Turner, and L. S. Flower, 1970. Bull. Wld. Hlth. Org. *42*:369-375.

Hansberry, R., 1968. Bull. Entomol. Soc. Amer. *14*:229-235.

Haskell, P. T., and A. J. Mordue, 1969. Ent. Exp. Appl. *12*: 591-610.

Haskell, P. T., and L. M. Schoonhoven, 1969. Ent. Exp. Appl. *12*:423-440.

Headley, J. C., and A. V. Kneese, 1969. Ann. New York Acad. Sci. *160*:30-39.

Henrick, C. A., G. B. Staal, and J. B. Siddall, 1973. J. Agr. Food Chem. *21*:354-359.

Hollingworth, R. M., 1971. Bull. Wld. Health Org. *44*:155-170.

Hollingworth, R. M., T. R. Fukuto, and R. L. Metcalf, 1967a. J. Agr. Food Chem. *15*:235-241.

Hollingworth, R. M., R. L. Metcalf, and T. R. Fukuto, 1967b. J. Agr. Food Chem. *15*:242-249.

Hori, M., K. Kakiki, S. Suzuki, and T. Misato, 1971. Agr. Biol. Chem. *35*:1280-1291.

Ishaaya, I., K. R. S. Ascher, and G. Shuval, 1974. Pestic. Biochem. Physiol. *4*:19-23.

Ishaaya, I. and J. E. Casida, 1974. Pestic. Biochem. Physiol. *4*:484-490.

Jackson, L. L., and G. L. Baker, 1970. Lipids. *35*:239-246.

Jackson, L. L., M. T. Armold, and F. E. Regnier, 1974. Insect Biochem. *4*:369-379.

Jacobson, M., 1972. *Insect Sex Pheromones*. Academic Press, New York.

Jermy, T., 1966. Entomol. Exp. Appl. *9*:1-12.

Jermy, T., 1971. Acta Phytopath. Acad. Sci. Hung. *6*:253-260.

Jermy, T., and G. Matolcsy, 1967. Acta Phytopath. Acad. Sci. Hung. *2*:219-224.

Johnson, O., 1972. Chem. Week. June 21 34-66; July 26 18-46.

Jumar, A., 1973. Z. Chemie. *13*:161-174.

Klun, J. A., C. L. Tipton, and T. A. Brindley, 1967. J. Econ. Entomol. *60*:1529-1533.

Koeppe, J. K., and R. R. Mills, 1972. J. Ins. Physiol. *18*: 465-469.

Lavie, D., M. K. Jain, and S. R. Sphan-Gabrielith, 1967. Chem. Commun. *1967*:910-911.

Lerner, A. B., and T. B. Fitzpatrick, 1950. Physiol Rev. *30*:91-126.

Lewallen, L. L., 1964. Mosquito News. *24*:43-45.

Lewis, D. K., 1967. Nature (London). *213*:205.

Locke, M., 1965. J. Ins. Physiol. *11*:641-658.

Loschiavo, S. R., 1969. J. Econ. Entomol. *62*:102-107.

Maddrell, S. H. P., 1972. New Scientist. *54*:203-205

Marx, J. L., 1973. Science. *181*:736-737.

Matolcsy, G., Gy. Saringer, R. Gaborjanyi, and T. Jermy, 1968. Acta Phytopath. Acad. Sci. Hung. *3*:275-277.

Matteson, J., H. M. Taft, and C. F. Rainwater, 1963. J. Econ. Entomol. *56*:189-192.

McFarlane, J. E., 1960. Can. J. Zool. *38*:231-241.

McMullen, R. D., 1959. Nature (London). *184*:1338.

McMullen, R. D., 1965. Can. Entomol. *97*:1200-1207.

Mehrotra, K. N., and A. Phokela, 1969. Arch. Int. Physiol. Biochem. *77*:799-806.

Menn, J. J., and M. Beroza, 1972. *Insect Juvenile Hormones; Chemistry and Action.* Academic Press, New York. 341 pp.

Metcalf, R. L., 1972a. J. Environ. Qual. *1*:10-14.

Metcalf, R. L., 1972b. *Implementing Practical Pest Management Strategies,* Proc. Nat. Extension Insect Pest Manag. Workshop, Purdue University. pp. 74-91.

Metcalf, R. L., 1972c. *Toxicology, Biodegradation and Efficacy of Livestock Pesticides,* (M. A. Khan and W. O. Huafe, eds.), Swets and Zeitlinger, Amsterdam, pp. 350-378.

Metcalf, R. A., and R. L. Metcalf, 1973. Pestic. Biochem. Physiol. *3*:149-159.

Miskus, R. P., T. L. Andrews, and M. Look, 1969. J. Agr. Food Chem. *17*:842-844.

Miskus, R. P., M. Look, T. L. Andrews, and R. L. Lyon, 1968. J. Agr. Food Chem. *16*:605-607.

Miyamoto, J., 1969. Residue Rev. *25*:251-262.

Mulder, R., and M. J. Gijswijt, 1973. Pestic. Sci. *4*:737-745.

Munakata, K., 1970. *Control of Insect Behavior by Natural Products,* (D. L. Wood, R. M. Silverstein, and M. Nakajima, eds.), Academic Press, New York, pp. 179-187.

Muniappan, R., 1972. J. Econ. Entomol. *65*:1482-1483.

National Academy of Sciences, 1972. *Principles of Plant and Animal Pest Control, vol. 3. Insect Pest Management and Control,* Publication 1695. U. S. National Academy of Sciences, Washington, D. C. pp. 283-285.

Norris, D. M., 1970. Ann. Entomol. Soc. Amer. *63*:476-478.

Norris, D. M., J. E. Baker, T. K. Borg, S. M. Ferkovich, and J. M. Rozental, 1970. Contrib. Boyce Thompson Inst. *24*:263-274.

Norris, D. M., S. M. Ferkovich, J. E. Baker, J. M. Rozental, and T. K. Borg, 1971. J. Ins. Physiol. *17*:85-97.

O'Brien, R. D., 1967. *Insecticides: Action and Metabolism,* Academic Press, New York. 332 pp.

O'Brien, R. D., L. Cheung, and E. C. Kimmel, 1965. J. Insect Physiol. *11*:1241-1246.

Ogita, Z., 1958. Botyu-Kagaku. *23*:188-205.

Persing, C. O., 1965. Bull. Entomol. Soc. Amer. *11*:72-74.

Pitman, R. M., 1971. Comp. Gen. Pharmacol. *2*:347-371.

Post, L. C., B. J. deJong, and W. R. Vincent, 1974. Pestic. Biochem. Physiol. *4*:473-483.

Preussmann, R., H. Druckrey, S. Ivankovic, and A. von Hodenberg, 1969. Ann. New York Acad. Sci. *163*:697-714.

Quraishi, M. S., 1971. J. Econ. Entomol. *64*:787-792.

Ripper, W. E., R. M. Greenslade, and G. S. Hartley, 1951. J. Econ. Entomol. *44*:448-459.

Robbins, W. E., J. N. Kaplanis, J. A. Svoboda, and M. J. Thompson, 1971. Ann. Rev. Entomol. *16*:53-72.

Rozental, J. M., and D. M. Norris, 1973. Nature (London). *244*:370-371.

Ruscoe, C. N. E., 1972. Nature (New Biol.). *236*:159-160.

Sacher, R. M., 1971a. Mosquito News. *31*:513-516.

Sacher, R. M., 1971b. Proc. 6th Brit. Insectic. Fungic. Conf. pp. 611-620.

Schaefer, C. H., and W. H. Wilder, 1970. J. Econ. Entomol. *63*:480-483.

Schaefer, C. H., and W. H. Wilder. 1972. J. Econ. Entomol. *65*:1066-1071.

Schneiderman, H. A., A. Krishnakumaran, P. J. Bryant, and F. Sehnal, 1970. Agr. Sci. Rev. *8*:13-25.

Schrader, G., 1961. Angew. Chem. *73*:331-334.

Schrader, G., 1965. World Rev. Pest Control. *4*:140-149.

Shorey, H. H., 1972. *Implementing Practical Pest Management Strategies*. Proc. Nat. Extension Insect Pest Management Workshop, Purdue University, pp. 63-68.

Shorey, H. H., 1973. Ann. Rev. Entomol. *18*:349-380.

Shorey, H. H., L. K. Gaston, and R. N. Jefferson, 1968. Adv. Pest Contr. Res. *8*:57-126.

Slama, K., 1971. Ann. Rev. Biochem. *40*:1079-1102.

Smissman, E. E., J. B. Lapidus, and S. D. Beck, 1957. J. Amer. Chem. Sco. *79*:4697-4698.

Spielman, A., and V. Skaff, 1967. J. Insect Physiol. *13*:1087-1095.

Sternburg, J., 1963. Ann. Rev. Entomol. *8*:19-38.

Sun, Y. P., and E. R. Johnson, 1972. J. Econ. Entomol. *65*:667-673.

Thomson-Hayward Chemical Co., 1973. Technical Information Bulletin, Kansas City, Kansas.

Thorsteinson, A. J., 1960. Ann. Rev. Entomol. *5*:193-218.

Turpin, F. T., and D. C. Peters, 1971. J. Econ. Entomol. *64*:1448-1451.

Twinn, D. C., 1972. Proc. 2nd Internat. IUPAC Cong. Pestic. Chem., Tel Aviv, 1970. (A. S. Tahori, ed), *1*:353-363.

Unterstenhofer, G., 1970. Pflanzenschutz-Nachrichten. *23*:264-272.

van Daalen, J. J., J. Meltzer, R. Mulder, and K. Wellinga, 1972. Naturwiss. *59*:312-313.

Vardanis, A., and L. G. Crawford, 1964. J. Econ. Entomol. *57*:136-139.

Walker, W. F., and W. S. Bowers, 1973. J. Agr. Food Chem. *21*:145-148.

Wallis, R. C., 1961. Mosquito News. *21*:187-189.

Weiden, M. H. J., 1971. Bull. Wld. Hlth. Org. *44*:203-213.

Wellinga, K., R. Mulder, and J. J. van Daalen, 1973. J. Agr. Food Chem. *21*:348-354.

Winston, P. W., and J. W. L. Beament, 1969. J. Exp. Biol. *50*:541-546.

Winteringham, F. P. W., 1965. *Aspects of Insect Biochemistry*, (T. W. Goodwin, ed.), Academic Press, New York, pp. 29-37.

Winteringham, F. P. W., 1969. Ann. Rev. Entomol. *14*:409-442.

Wood, D. L., R. M. Silverstein, and M. Nakajima, 1970. *Control of Insect Behavior by Natural Products*, Academic Press, New York.

Wright, D. P., Jr., 1963. Adv. Chem. *41*:56-63.

Wright, D. P., Jr., 1967. *Pest Control: Biological, Physical and Selected Chemical Methods*. (W. W. Kilgore and R. L. Doutt, eds.), Academic Press, New York, pp. 287-293.

Chapter 6

TARGET SPECIFIC PESTICIDES: AN INDUSTRIAL CASE HISTORY

Gustave K. Kohn*
Ortho Division
Research and Development Department
Chevron Chemical Company
Richmond, California

This chapter deals with the history of an attempt to develop a target-specific insecticide. It will include the genesis of the project, certain of the chemical and biological aspects, and some economic and business considerations. For the portion of this chapter that is "transscientific" {Weinberg, 1972} in nature the author takes personal and exclusive responsibility.

Let us examine what is meant by target-specific pesticide. The term is a somewhat general abstraction. As such, it may refer to a chemical which interacts only with a single organism, or more fundamentally, with a highly specific biochemical system possessed by that organism alone. More frequently, a target-specific pesticide refers more to use pattern than to biochemical specificity. That is, the substance is only applied against one

*Present affiliation: United Nations Industrial Development Organization, New Delhi, India.

or a small group of related organisms against which it is highly effective, being relatively innocuous to other organisms, generally environmentally linked to the target species. The modified insecticidal carbamate RE-11,775 is a target-specific pesticide, particularly in this latter sense.

We have emphasized that the term target-specific pesticide is an idealized abstraction. With the discovery of ecologically undesirable consequences deriving from the use of broad-spectrum persistent pesticides, beginning with DDT, many concerned scientists felt that a chemical that would only affect a given organism would be more environmentally sound and desirable.

Those of us with some biochemical experience realize that the above conception, however desirable, is an unattainable ideal toward which we may strive but never wholly succeed. Living species possess at least as much, actually more, similarity to each other than dissimilarity. The principle of chemotherapy, and chemical pesticides are a highly useful branch of chemotherapeutics, is not complete target specificity but differential toxicity. Stated in its simplest form, a useful pesticide is more effective against one or more undesirable organisms than it is against the large number of other organisms (host or otherwise) with which the pesticidal agent must inevitably come into contact.

An authority on medical chemotherapeutics {Albert, 1960} writes, "Selective toxicity is the injury of one species of living matter without harming another species with which the first is in intimate contact. The species which is to be injured is the uneconomic species and the species which is to be preserved is the economic species."

We prefer the term differential toxicity to avoid the implication of absolute selectivity. Most often there is a quantitative distinction rather than an absolute qualitative difference.

We live in an era of examination of the alternate strategies of pest control and crop production. It is necessary, then, that we understand the meaning, the limitations, and the potentials associated with the target-specific pesticides, since the development of target-specific pesticides has been strongly recommended as one of the better strategies for pest control in the future.

We did not try to develop a target-specific pesticide. RE-11,775 derived from a program of modifying insecticidal carbamates by chemical alteration of parent molecules.

At an ACS Symposium on new carbamate insecticides about 10 years ago, we described and subsequently published on a group of simple alkyl phenyl carbamates {Kohn et al., 1965; Moore, et al., 1959}. Fig. 1 provides the general formula for that group. In analyzing the biological data we found maximum activity when R_1 was methyl, R_2 was H, and R_3 was a secondary alkyl.

Activity was highest when R_3 was meta, was somewhat less when R_3 was ortho, and the structure was practically inactive when R_3 was para. Further, the biological activity was variable over the spectrum of insect species examined. Finally, maximum useful insecticidal activity was discovered where R_3 was 4 and 5 carbons falling off along a smooth curve as the chain length was increased or decreased. A surprising result was that secondary alkyl was generally more active than tertiary alkyl while normal or iso-substituted alkyls possessed relatively much less anticholinesterase effect.

A biological property that limited the usefulness of these compounds was their relatively high mammalian toxicity for the most active insecticidal compounds. Despite this limitation one of these compounds, particularly as a dilute granular formulation, has been a highly commercially successful corn rootworm pesticide for the past several years. Fig. 2 gives the

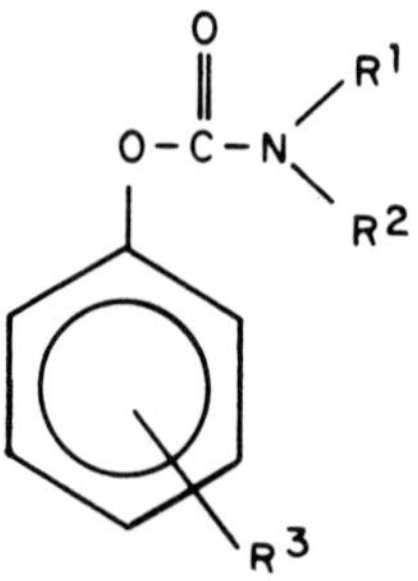

FIG. 1. Simple alkyl-substituted phenyl carbamates. R^1= H or lower alkyl. $R^2 = CH_3$. R^3 = primary, secondary, and tertiary alkyl; also, iso and unsaturates. {From Kohn, et al., J. Agri. and Food Chem. 13:232, 1965}.

structure of BUX developed from this early program. An extremely simple alkylation process {Kohn and Stevick, 1968} provides the procedure to produce the high meta content, and the proportion of 1-methylbutyl to 1-ethylpropyl isomer given in the figures derives from that process.

In more recent years we have been looking at modifications of these molecules that would on the one hand increase the toxicity index--defined here as the ratio of: $\frac{LD_{50} \text{ insect(s)}}{LD_{50} \text{ oral, rats}}$

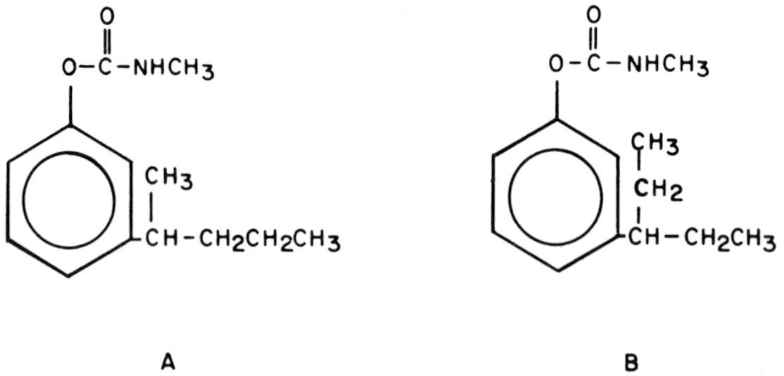

A

B

FIG. 2. BUX insecticide. Major isomer 65% min. A:B=3:1. {From Ospenson et al., 1959; Kohn et al., 1959}.

and on the other hand extend either the scope or the specificity within this series. The subject pesticide RE-11,775 {Brown and Kohn, 1969} given (Fig. 3) derives from that program. Although we consciously tried to alter the biological activity, those alterations that we eventually discovered were not anticipated nor are the reasons for the resulting specificities thoroughly understood.

PREPARATION OF N-SULFENYLATED METHYL CARBAMATES

Starting with the simple secondary alkylphenyl N-methyl carbamates various nitrogen-substituted derivatives were prepared. This chapter deals solely with the N-sulfenylated compounds. Initially the carbamate was reacted with an alkyl or aryl sulfenyl halide generally in the presence of a base and usually in a convenient solvent {Brown and Kohn, 1969}. Fig. 4 describes this preparation method.

A wide variety of solvents are permissable. Tertiary amines are preferred as the acceptors. The reactions are generally run at ambient pressures and temperature from 0--60°C.

Since this initial work, other methods for obtaining N-sulfenylated products have been developed which are the subject

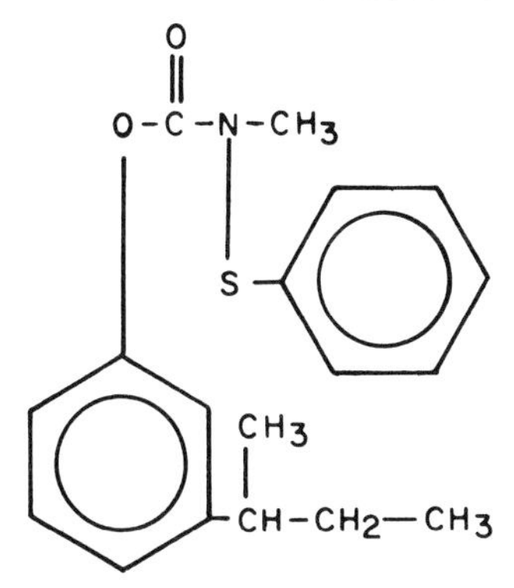

FIG. 3. RE-11,775, major isomer. {From Brown and Kohn, 1969}.

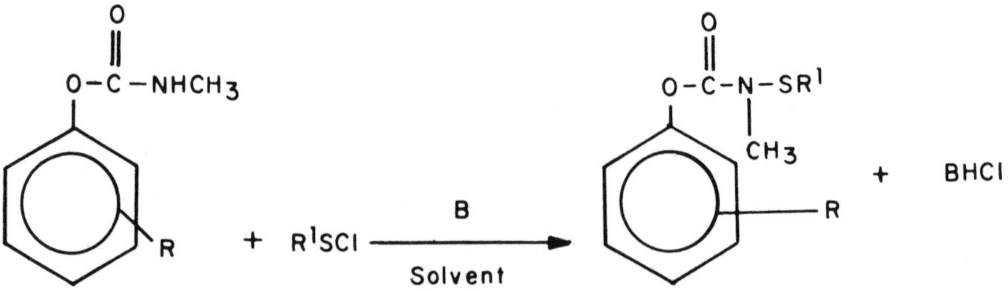

FIG. 4. Synthesis of RE-11,775 group

of pending patents. Some of these avoid the need for the preformed carbamate, but the previous method as described above is simple and easy to perform and is essentially very useful where manufacturing know-how and facilities already exist for the carbamate intermediate. This becomes an important factor when considering the economics of a target-specific pesticide.

The sulfenylated methyl carbamates derived from the m-secondary butyl and m-secondary amyl phenols provided some of the biologically more interesting compounds. We chose for preliminary development the compound given in Fig. 3. For this preparation the benzene sulfenyl halide is simply prepared by known literature methods either from the reduction of the readily available sulfonyl chloride or from the reduction of the disulfide.

METHODS

The methods of synthesis of the secondary butyl and amyl N-methyl carbamates and the sulfenylated derivatives therefrom are discussed in Kohn et al., {1965}; Moore et al., {1959}; Kohn and Stevick, {1968}; Brown and Kohn, {1969}. Although other methods are also described the practical synthesis consisted of alkylation of phenol followed by catalytic isomerization {Kohn and Stevick, 1968} and then by direct reaction of benzene sulfenyl chloride on the appropriate N-methyl carbamate.

The Chevron biological mosquito assays are given below.

Testing

Dixie cups are filled with 100 ml tap water each containing 10 fourth stage larvae or pupae. One pellet of Purina Rabbit Checkers is added to each cup as larval food. Other stages of larvae may be used for special testing. Three replicates per compound are employed, each test container being properly labeled on the upper inside rim of the cup.

For initial screening and follow-up testing toxicant solutions are delivered to the water in the cup containing the test organisms by using 5 lambda disposable micro-pipetts ("Microcaps").

Immediately after treatment, the water is thoroughly mixed using a plastic pot label. Each test cup is then covered with the bottom of a plastic petri dish to prevent the escape of emerging adults and to prevent evaporation.

Test units are held at room temperature (75°F) for 24 hours.

Evaluation

Mortality counts are made after the larvae or pupae have been exposed to the treated water for 24 hours.* The percent control is calculated as the average percent insect mortality based on an exact mortality count of the insects in each replicate. Moribund larvae unable to rise to the surface after being forced to submerge or after being touched are classed as dead.

The procedure for the laboratory and field observation cited from Dr. Schaefer's work at the Mosquito Control Labora-

---

*If desired, delayed effects can be determined by holding the treatments until adult emergence (3-4 days). At that time evaluation is made for pupal and adult mortality and any obvious morphological aberrations are noted.

tories at Fresno are summarized below. Some are given later in the discussion. The adult mosquitoes were reared from larvae collected in the field from areas of known organophosphorus resistance or susceptibility. Adults were tested by a 1-hour exposure to filter paper treated with a given quantity of toxicant in an acetone solution, after which they were held in untreated containers and evaluated for mortality 24 hours after exposure. Each treatment was replicated at least six times and each replication consisted of 20 adults.

BIOLOGICAL PROPERTIES OF RE-11,775

In most important areas of economic insect control sulfenylation of the carbamate moiety provided little encouragement. Both the meta secondary-butyl and the meta secondary-amyl methylcarbamates had many areas of relatively high activity. They were both weak on many adult lepidopteran and leaf-feeding larval forms. Sulfenylation only added to the ultimate complexity of manufacture but did not improve biological activity. In some cases, the activity was markedly reduced.

For example, Table 1 illustrates a marked reduction in biological activity resulting from the sulfenylation of RE-5305, the N-methylcarbamate derived from m-sec.butylphenol. RE-5305 initially showed considerable promise for cockroach control. Both the American and German species, either by topical application or by the more practically significant exposure to residues technique, were less susceptible to the sulfenylated product. Such reductions were unusual for in general the activities of the carbamates and their sulfenylated derivatives were similar.

Tables 2 and 3 provide the data which aroused some interest. In these tests water containing alfalfa meal is used so that the medium contains an active microorganism population and approximates field ponds.

TABLE 1

Activity of RE-11775 with German Cockroach (G) (Blattella germanica) and American Cockroach (A) (Periplaneta americana)

Deposit, mg/ft^2	Percent control 24 hr RE-5305		Percent control 24 hr RE-11775	
	A	G	A	G
100	100	100	96	0
30	100	76	66	0
10	100	33	23	6
3	100	10	16	3

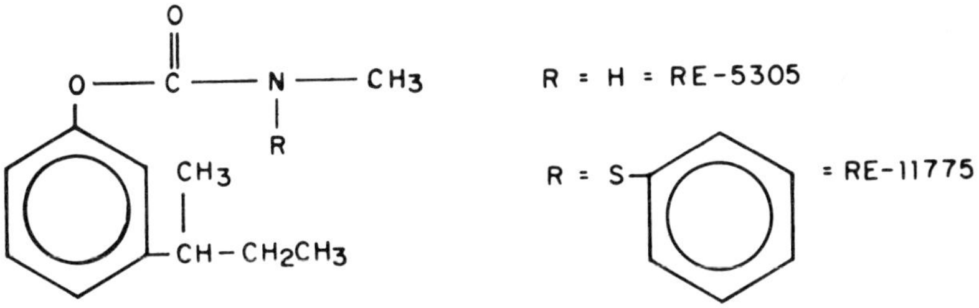

O—C(=O)—N(R)—CH_3 on a phenyl ring bearing $CH(CH_3)-CH_2CH_3$

R = H = RE-5305

R = S-phenyl = RE-11775

The mammalian toxicity of RE 5305 had provided a limitation to its development. A reduction of this toxicity by sulfenylation would be most desirable. Table 4 provides the results for one such set of values. This provided a toxicity reduc-

TABLE 2

Insecticidal Activity of m-sec. butylphenyl-N-methyl Carbamates with Aedes aegypti Larvae

Conc. in H_2O, ppm	Percent control	
	RE-5305	RE-11775
0.310	100	100
0.125	96	100
0.050	0	90

TABLE 3

Activities of RE-11775 and Reference Insecticides Tested with *Aedes aegypti* Pupae

Conc. in H_2O	Percent control RE-11775	Baytex[a]	Abate[b]
0.1	80	0	2.5

[a]Baytex (Bayer)
[b]Abate (Amer. Cyan.)

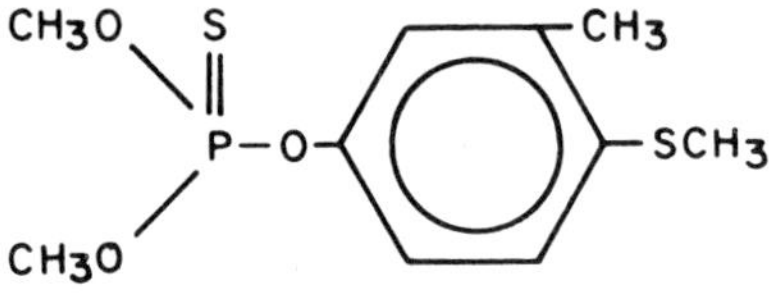

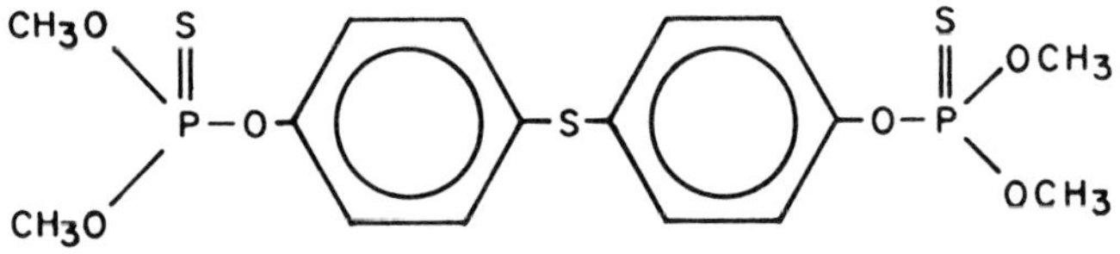

tion of sixfold which when combined with mosquito toxicity data encouraged further developmental work.

TABLE 4

Mammalian Toxicity (Oral) of m-sec. Butylphenyl-N-methyl Carbamates

	A RE-5305	B RE-11775
LD_{50} rats, mg/kg	16	100+

*Toxicity index $\frac{B}{A}$ = 6+

These interesting leads in the area of mosquito control were then pursued further. Samples of RE-11,775 were tested at the Mosquito Control Research Laboratory, University of California, Fresno, California, under the direction of Dr. Charles Schaefer {Schaefer and Wilder, 1970; Schaefer, 1972; Schaefer and Dupras, 1972}.

A very particular and serious problem exists in the San Joaquin Valley. During the rainy season and after irrigation, much of the grazing land is partially submerged. The cattle grazing in this area leave depressions as they walk through the muddy fields. These fill with water, and with the naturally submerged areas, the irrigation ditches, etc., there is provided an ideal breeding place for certain varieties of mosquitoes. One of these, Culex tarsalis, is the important local vector for a viral encephalitis which becomes endemic in California about once every 3 or 4 years. Other Culex and Aedes species are also involved as vectors. The disease affects man and horses and occurs in Texas and other areas of the southwest as well as California. Larvae of this mosquito are highly resistant to most organophosphates and chlorinated hydrocarbon and some carbamate insecticides. A definite need presently exists for the control of these important disease vectors.

Laboratory and small-scale field tests corroborated the insecticidal activity of RE-11,775 discovered in our laboratories. Further, RE-11,775 provided good control even where _resistance to conventional pesticides was high_. The next tables are quoted with permission from work performed by Dr. Schaefer of the University of California.

BUX, RE-11,775, and Parathion were evaluated in the laboratory against two species of mosquito larvae, _Culex pipiens quinquefasciatus_ and _Aedes nigromaculis_ (phosphate resistant and nonresistant strains). The _A. nigromaculis_ were field collected and the C. pipiens quinquefasciatus were from a laboratory colony. All data are based on mortality at 24 hours. The test was not replicated (Table 5).

In the following laboratory test, concentration series were made and fourth instar Culex pipiens quinquefasciatus larvae were tested for mortality at various holding temperatures. Each treatment was replicated three times, each replicate consisted of 20 larvae per cup (Table 6).

All of the Culex tarsalis larvae used in these tests were field collected from specific areas of known organic phosphate resistance. Larvae were tested in waxed-paper cups by treating the water with 0.001 percent toxicant in acetone. Each replication consisted of 20 larvae in a cup and each treatment was

TABLE 5

Effectiveness of RE-11775 in Laboratory Larval Tests on Field-collected Mosquito Strains

	LC_{50}, ppm	LC_{90}, ppm
Culex pipiens quinquefasciatus		
RE-11775	0.003	0.0043
BUX	0.047	0.069
Parathion	0.028	0.004
Aedes nigromaculis (susceptible strain)		
RE-11775	0.0048	0.0079
BUX	0.037	0.064
Parathion	0.0035	0.010
Aedes nigromaculis (resistant strain)		
RE-11775	0.010	0.025
BUX	0.12	-----
Parathion	0.23	0.58

TABLE 6

RE-11775 Activity: Effect of Temperature on *Culex pipiens quinquefasciatus* Larval Control (Laboratory Test)

Treatment	Temperature	ppm of RE-11775 required for LC_{50} and LC_{90}	
		LC_{50}	LC_{90}
1. 11775	$60^{o}F$	0.0027	0.0043
2. 11775	$70^{o}F$	0.0026	0.0041
3. 11775	$80^{o}F$	0.0029	0.0044

replicated twice. All larvae mortality is based on 24-hour evaluation (Table 7).

A comparison of the incapacity of test mosquito larvae to withstand RE-11,775 is given in Table 8. In this the toxicity index, defined as $\frac{LD_{50} \text{ resistant strain}}{LD_{50} \text{ susceptible strain}}$, is just slightly higher than one for RE-11,775 while it is 666 for Parathion.

TABLE 7

RE-11775 Control of Organophosphate-resistent *Culex tarsalis* Larvae

Resistant strain location	Test dates	LC_{50}, ppm	LC_{90},* ppm
1. Caruthers (Fresno County)	10/27/69	0.003	0.0078
2. Malaga (Kings County)	10/20/69	0.0038	0.0088
3. James (Kern County)	10/13/69	0.0035	0.0070
4. Jessups (Kern County)	10/13/69 10/18/69	0.0042 0.0040	0.0092 0.0080

*Small difference between LC_{50} and LC_{90} is significant.

TABLE 8

Insecticidal Toxicities with Susceptible and Resistent *Aedes nigromaculis* (adult)

Compound	Susceptible LC_{50}	Susceptible LC_{90}	Resistant LC_{50}	Resistant LC_{90}	$LC_{50}R \div LC_{50}S$
Parathion	0.017	0.049	7.8	22.2	666
Fenthion (Baytex)	0.0068	0.016	0.15	0.35	22
Dursban	0.0033	0.0048	0.15	0.31	5
DIBROM	0.0054	0.0094	0.062	0.16	11
BUX	0.019	0.032	0.026	0.048	1.4
RE-11,775	0.009	0.014	0.015	0.037	1.6

Other well-known mosquito larvacides reveal weakness in the control of the resistant strains.

The special 4 lb/gal formulation was flown on an 80-acre rice field at 3.2 oz formulated material per acre. The *Culex tarsalis* in this field were OP-resistant, uncontrolled with Methyl Parathion and Fenthion, but completely controlled with RE-11,775 (Table 9).

TABLE 9

RE-11775 Control of *Culex tarsalis* Larvae in Rice Paddy Employing ULV Air Application

Treatment	Formulation/A active/acre	Result of UVL air application, percent control: 6 hr	24 hr
11,775	3.2 oz/0.1 lb	98	100

*Previous efforts to control this population with 0.1 lb methyl parathion and 0.2 lb Fenthion gave little or no control.

FURTHER DEVELOPMENT WORK ON RE-11,775

Having manufactured many millions of pounds of BUX Insecticide, process development work mainly consisted of studies related to:

1. Acquisition of adequate supplies of butene rather than pentene.
2. Studies of needed modifications of manufacturing and distillation equipment for the handling of the smaller hydrocarbon in the various process steps.
3. Studies relating to the sulfenylating of the preformed carbamate.

It is not our intention to review these studies in detail here except to say that there were no foreseeable obstacles for the preparation of RE-11,775 from a process standpoint. Some minor equipment and operating procedural changes would be required in adapting equipment from BUX to RE-11,775 manufacture. The pesticide could be produced and the product cost would be in the middle range for ordinary pesticides.

Coincident with these studies, grants-in-aid were established at U.C. Berkeley, under Prof. John Casida and under Prof. Francis Gunther, U.C., Riverside, to look at animal metabolism and analytical methodology, respectively. An analytical procedure that would be employed for residue determination was developed. This method consists of an extraction and VPC analysis. The metabolism of RE-11,775 was studied from ring-labeled, chain-labeled, and methyl-labeled ^{14}C. This study has been published in the Journal of Food and Agricultural Chemistry {Cheng and Casida, 1973} and will not be quoted here except insofaras to say that the metabolites were those consistent with previous studies on carbamate insecticides by Professor Casida and his students. There were no unexpected intoxifications from the metabolites.

Though much more work would have to be done, initial experimental results on toxicological effects and ecological impact indicated no extraordinary problems.

ECONOMIC CONSIDERATIONS

At this point, we have reviewed the content that usually relates to pesticide science. We have covered synthesis, process, biological evaluation, analytical methods, metabolic aspects, although quite sketchily. The areas next covered relate to, but transcend science, and the reader should not apply to the numbers and ideas expressed below the same critical standards that he employs toward the matter in the first part of this chapter.

We will now look at market potential and decision-making procedures, both by industry and government. It is, at best, quite difficult to ascertain the market potential of pesticide chemicals. For a mosquito larvacide the uncertainties multiply. For 1969 our company experts ascertained that the total California market was somewhat less than 250,000 lb. This was based on the total amount of chemicals exclusive of oil sprays that were utilized for mosquito control in the State. Slightly less than 400,000 gal of oil were similarly employed.

For the rest of the country and the world the emphasis is on adult control rather than larvaciding. Perhaps 1 million lb of a specific larvacide is a fair world market estimate. Almost all of the WHO effort quite reasonably stresses adult control. For these uses (disease vector) programs, larvaciding has not progressed beyond the screening and research stage.

The total market then appeared to be about 1 million to a 1-1½ million lb. However, no one chemical ever attains the total market and it is not unreasonable then to base the economics for development of this target-specific chemical at 750 thousand lb ultimate sale. Even this was a highly optimistic projection.

Blair {1969} estimated the cost for development of a pesticide at $4.1 million. Both inflation and very much additional regulation agency requirements raises that number to $7 million. These figures of course refer to broad-spectrum pesticides.

Nevertheless, if a chemical is applied to grazing areas, to irrigation ditches and sources of irrigation, and to potable water a host of analytical toxicological and ecological questions will be asked and must be answered. Water quality control is centered presently under EPA. The questions answered must satisfy the protocols of that agency and the research cost for the determination of the answers is not inconsiderable.

Financial statements are generally more noteworthy for what they conceal rather than for what they reveal.

We cannot go into all of the economic detail. Economic detail of this type would reflect this corporation's accounting procedures, would require pages of explanation, and is not directly transferable as knowledge to experience of most scientists whose professional attachments are elsewhere. We will briefly summarize the story of the conclusion of this project.

SUBSEQUENT CASE HISTORY OF RE-11,775

After an examination of the market potential, the cost of manufacture, and the return Chevron felt that although this chemical might prove useful as a larvacide--particularly against the vectors of viral encephalitis--the company could not proceed with its development. The metabolic work and the analytical research had exposed no obviously discernible obstacles. We felt that the environmental considerations also would provide no bar. Even the most favorable projections showed a loss to Chevron if it pursued the development of this chemical.

We then spoke with officials of the state of California with, in effect, the following proposition. Chevron would donate to the State all the rights under the various compositions and process patents for manufacture by the State. As an alternate we agreed to manufacture the chemical on a cost plus basis.

It should be emphasized here that both the State and the company acted in good faith and, as far as this author can see, beyond reproach. From the standpoint of the State, mechanisms did not and do not exist that would permit, without legislative action, the offers submitted by Chevron. On the other hand, Chevron, whose research is quite frankly and unashamedly aimed at producing chemicals for a profit, could not proceed with further expenditures with an unprofitable, really a losing, project. Again there is no blame that can be justifiably leveled at any of the parties for this unsuccessful development, except perhaps at the chemists who should have produced a more perfect product.

CONCLUSIONS AND REMARKS

It is proper to say that no mechanisms exist in our society, at present, for the development of industrially unprofitable target-specific pesticides. Does this apply then to all target-specific pesticides? This cannot be answered unequivocably yes or no. The following are some propositions which we present in conclusion about the possibilities of development of target-specific pesticides for the near future. These propositions are based upon the premise of no large-scale alteration of the nature of American agriculture and its organization or the economic system as a whole, again for the near future. These propositions represent this author's reflections exclusively.

1. No chemical is an absolute target-specific pesticide.
2. If the target is specific for a broad, agriculturally significant organism, or public health pest as, for example, the boll weevil, the bollworm, Johnson grass, nutsedge,

Phytophora infestans, Ventura inequalis, corn rootworm, grain smut, rats, etc., then the chances for successful development are at a maximum. Large markets are foreseeably involved.

3. As a corrolary of Item 2, if the target organism is too specific, e.g., a specific race or variety of an organism, or an organism of local significance only, and, particularly for an organism for which the total market for control is small, the opportunities for development diminish accordingly and are all but impossible.
4. Even for target-specific chemicals of low or intermediate market potential, the flexibility of our socioeconomic system could be made to accept these chemicals. There are two main routes for an accommodation. (1) The federal government and the states could relax certain requirements for registration and hence reduce the total development cost. (2) Local state and federal authorities could encourage production and distribution of the target-specific chemical by some form of subsidy. Such subsidy could take many forms. Cost-plus production, sharing of developmental costs, state production, tax rebate, etc.
5. If a social good exists then the democratic process can be made to move by one of a variety of means to the acquisition of that good. A broad view of what the economist calls the externalities, both advantages and disadvantages, relating to such target-specific pesticides is the subject of considerable contemporary exploration in the current economic literature. Large government subsidized research projects such as the present boll weevil attractant sterilization program, and others like it, will ultimately lead to the accommodations discussed in Item 4.

In this chapter, we have outlined the case history of a target-specific pesticide. The meaning of the term target-specific has been briefly explored. The chemistry and biological activity of the pesticide were outlined. From this case history we have examined the possibilities for future development of target-specific pesticides pointing toward those factors which would lead to the successful development and those that would hinder such development.

For the cases of a target-specific pesticide of small market potential but for which there is genuine social need, we have argued for the creation of mechanisms within the framework of our political system for their development. We have, in addition, outlined what we consider to be viable courses of action. Within this outlined framework the author foresees the development within the next two decades of certain target-specific pesticides which render to society assets of indisputable social value.

REFERENCES

Albert, Adrian, 1960. *Selective Toxicity*, John Wiley & Sons, New York, p. 1.

Blair, E. H., 1972 Chem. & Eng. News *50*(17):20.

Brown, M. S., and G. K. Kohn, 1969. N-Substituted Aryl Carbamates, U.S. Pat. 3,663,594.

Cheng, Hong-Ming and John E. Casida, 1973. J. Agr. Food Chem. *21*:1037.

Kohn, G. K., J. N. Ospenson and J. E. Moore, 1965. J. Agr. Food Chem. *13*:232.

Kohn, G. K. and L. Stevick, 1959. Insecticidal Carbamates, U.S. Pat. 3,062,867.

Kohn, G. K. and L. Stevick, 1968. Isomerization Process, U.S. Pat. 3,655,780.

Moore, J. E., J. N. Ospenson, and G. K. Kohn, 1959. m-sec. Butylphenyl-N-Methyl Carbamates, U.S. Pat. 3,062,865.

Schaefer, C. H., 1972. Proc. 40th Annual Conf. of the Amer. Mosquito Control Ass'n., Miami Beach, August 23, 1972.

Schaefer, C. H. and E. F. Dupras, 1972. Mosquito News *32*:201

Schaefer, C. H. and W. H. Wilder, 1970. J. Econ. Ent. *63*:480

Weinberg, A. Win, 1972. Science *177*:211

Chapter 7

COMPARATIVE BIOTRANSFORMATION AS A FORECASTER OF THE ECOLOGICAL CONSEQUENCES OF SELECTIVE INSECTICIDES

Tsutomu Nakatsugawa and Robert Stewart
Department of Entomology
State University of New York
College of Environmental Science and Forestry
Syracuse, New York

Although selective toxicity has long been a major issue of insecticide toxicology, even the broad selectivity between insects and mammals has been an elusive goal to achieve {Hollingworth, 1971}. In the absence of perfect insecticides with ideal selectivity, we have to take calculated risks by weighing them against the benefits. On the other hand, the recent recognition of the impact of insecticides on the ecosystem has led to increasingly stringent requirements for insecticide use. Thus, forecasting undesirable effects of insecticides on nontarget organisms is a necessity difficult to fulfill.

It is now obvious, however, that a mere survey of insecticide residues and population changes in fauna and flora after a test application of an insecticide falls far short of that goal. Although assessment of acute lethal effects on nontarget species (imperfect selective toxicity) is an important part of the risk evaluation, there are other obvious toxicological side effects that need to be considered. These risks may arise directly from the application of a chemical or may become apparent only through complex chemical-organism interactions in the ecosystem (distant effect). A full consideration should include purely ecological consequences resulting from primary disturbances of the ecological balance as well as the fate of toxicants in plant, microbial, and nonbiological systems. We shall limit our discussion, however, to the physiological aspects of the problem and first consider the biochemical basis of various undesirable effects of insecticides on nontarget species. We shall then survey the current potential of risk evaluation for animal species based on our knowledge of biotransformation.

PRIMARY EFFECTS ON NON-TARGET SPECIES; IMPERFECT SELECTIVE TOXICITY

Since insecticides have been screened for their high lethal activity at a low dose, they all have high affinity to or reactivity with a primary biochemical target that is vital to the life process (biochemical selectivity). If the target entity is limited to a certain group of animals, the insecticide would probably be highly selective against those species. Such targets have been suggested {O'Brien, 1967}, e.g., noncholinergic neuromuscular junction, α-glycerophosphate shuttle, chitin and trehalose metabolism, and peripheral inhibitory neurons of insects. Recently developed insecticides that seem to interfere with chitin synthesis {Post and Vincent, 1973} as well as insect hormone mimics fall in this category. The majority of current insecticides, however, involves targets that are more or less a part of the general biochemical mechanism common to a wide variety of life forms. Thus, the fairly high biochemical selectivity due to specificity of targets does not find its expression at the interphylum or interspecies level.

FAULTY BIOCHEMICAL SELECTIVITY, THE MOLECULAR BASIS OF PHYSIOLOGICAL SIDE EFFECTS

Nontarget species are also affected by attack of a biochemical target unrelated to the primary lethal action of the toxicant. Although biochemical mechanisms are not always understood, these effects may be conveniently categorized under secondary and tertiary effects.

Under secondary effects we may classify those effects that are due to the latitude of the lethal mechanism. Biomacromolecules or functional units that are sufficiently similar to the primary biochemical target of lethal action can interact with the toxicant by the same molecular mechanism as in the primary action.

An insecticide molecule may possess more than one toxicological property with different mechanisms of action. Such an additional property may cause another type of toxicant-target interaction which we might call tertiary effects. Although bifunctional insecticides have been suggested {Holan, 1971; Metcalf et al., 1966; Wilkinson, 1971}, tertiary effects usually result from an unintentional peculiarity of the molecules. Also, tertiary effects may arise from some of the degradation products of insecticides. Most teratogenicity, mutagenicity, and carcinogenicity are tertiary effects. Secondary or tertiary interactions do not necessarily result in the side effect, however, these effects occur when the extent of interactions is significant enough and the target is vital to the well-being of the species concerned.

The known side effects are practically limited to chlorinated compounds and anticholinesterase insecticides {Pimentel, 1971}. Perhaps the best examples of the secondary effect are delayed neurotoxicity {Johnson, 1969; Aldridge et al., 1969} and teratogenicity {Flockhart and Casida, 1972; Roger et al., 1969} of some organophosphates. Available evidence indicates that these effects are caused by a single dose of the chemicals and are the result of the typical S_N2 reaction between the organophosphates and the serine hydroxyl (phosphorylation) of a hypothetical esterase that is critical for maintenance of nerve cells {Johnson, 1969} and membrane transport {Roger et al., 1969}, respectively.

Organophosphates, particularly dimethyl esters, are good alkylating agents {Hilgetag and Teichmann, 1965; Eto and Ohkawa, 1970}. Since DNA is a well-known target of alkylating agents, mutations and chromosomal aberrations caused by dichlorovos in plants and bacteria may be a tertiary effect related to alkylating capacity of this compound {Fishbein et al., 1970}. In vitro alkylation of DNA bases by organophosphates has been reported {Lofroth, 1970}. The relation of the effects of organophosphates to the molecular reactivity is illustrated in Fig. 1.

Side effects of chlorinated insecticides are more difficult to define mainly because of our lack of knowledge of the primary molecular mechanism of action of these compounds (Fig. 2). While the exact nature of the interaction is unknown, the primary

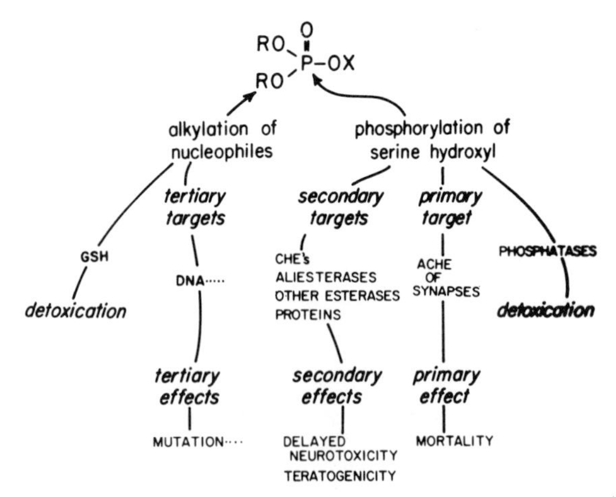

FIG. 1. The reactivity and toxicological effects of organophosphates.

Primary target ------- Ionic balance mechanism of critical nerve cells
sodium-gate, ATPase, electron acceptor, ...?

Secondary targets ----- Ionic balance of other cells

Tertiary targets -------------- ?

Secondary effects ----- Toxicity to some aquatic organisms

Other side effects ----- Enzyme induction
Reproductive failure
Phytotoxicity
Behavior change

FIG. 2. Action of DDT and related compounds.

target of DDT's insecticidal action is probably in the nerve membrane and is vital to ionic balance of the nerve cell, whether it be the hypothetical "sodium-gate" {Hilton et al., 1973}, ATPases {Koch, 1969; Matsumura et al., 1969; Matsumura and Patil, 1969} or a macromolecular electronacceptor {Matsumura and O'Brien, 1966; Wilson et al., 1971}. Peculiar susceptibility of fish and other aquatic forms to DDT and related compounds may be due to a secondary effect, i.e., the inteference with the ionic balance {Janicki and Kinter, 1971; Matsumura, 1972}. The possible relation of similar action of chlorinated insecticides to enzyme induction has also been suggested {Wang and Matsumura, 1969}, although the induction could be a tertiary effect. Reproductive failure of certain bird species may be related to various factors depending on the species affected {Switzer and Lewin, 1971}. The cause may be behavior change of adult birds {Switzer and Lewin, 1971} or eggshell thinning {Heath et al., 1969; Wiemeyer and Porter, 1970; Longcore et al., 1971}, which in turn may be related to estrogenic activity of some of the chlorinated compounds {Bitman et al., 1968; Cecil et al., 1971; Gellert et al., 1972}. Although inhibition of carbonic anhydrase by DDT {Torda and Wolff, 1949} has been linked to the eggshell thinning {Bitman et al., 1969}, the inhibition may not be a toxicological property of DDT {Pocker et al., 1971}. Interference with thermal acclimation mechanisms {Anderson, 1971} by DDT may also contribute to damage of some fish populations. The relation of the reported inhibition of photosynthetic electron transport by DDT {Rogers et al., 1971} to other actions of DDT is not clear.

Although no side effects have been noted in the use of nicotinoids, secondary biochemical targets (esterases) might be worth further attention in view of their effects on acetylcholinesterases {Yamamoto et al., 1968}. Although rotenoids seem to have a very strict biochemical selectivity at a low dose even to allow "titration" of the specific target {Ernster et al., 1963}, possibly a nonheme iron protein on the NADH side of coenzyme Q {Bois and Estabrook, 1969}, a wider target latitude is revealed at higher concentrations {Horgan et al., 1968; Teeter et al., 1969} as indicated in Fig. 3. If their ecological safety is to be assured, effects on other electron transport systems by rotenoids and their metabolites may deserve further investigation. A number of secondary and tertiary targets have been noted in the course of the study of the lethal action of insecticides, but have mostly been dismissed as unimportant. These effects may deserve more serious attention as potential mechanisms of side effects.

Only the obvious and acute effects, however, will be diminished by the use of selective insecticides. Yet, experience and basic information accumulated to date might furnish us with

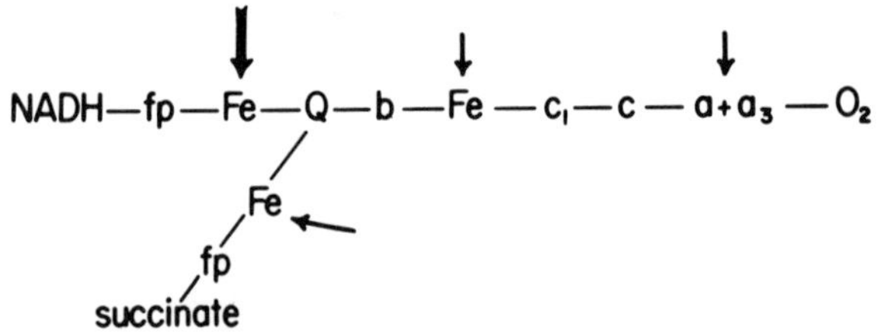

NADH—fp—Fe—Q—b—Fe—c_1—c—$a+a_3$—O_2
Fe
fp
succinate

FIG. 3. Targets of rotenone in mitochondria.

enough guidelines to foresee other side effects. It is obvious from the foregoing discussion that basic knowledge of insecticide action at the enzymatic and molecular level is essential for successful forecasting of the impact of insecticide use.

ASSESSMENT OF BIODEGRADATION POTENTIAL AND ITS IMPLICATIONS

Although nontarget species can be affected immediately after the use of insecticides or at a distance (in time or space), the current shift from the so-called hard insecticides to biodegradable types was prompted by distant effects, i.e., the measurable damage of nontarget species through biomagnification of some chlorinated insecticides of low biodegradability and by the ubiquitous presence of residues of these chemicals or their terminal residues in nontarget species including man. Manifestation of distant effects is subject to the sensitivity of the species and a number of factors that control the concentration of the toxicant at the biochemical target. Biodegradation is probably the single most important factor of all since it influences the input as well as the elimination of toxicants for many species. The majority of insecticide biotransformations in animals follow surprisingly few kinds of pathways, i.e., oxidation by microsomal oxidases, nucleophilic cleavage by glutathione S-transferases and hydrolases. While the latter two types of enzymes handle only certain groups of insecticides, almost all insecticides are potential substrates of microsomal oxidases. Thus, information on these oxidases offers the most important common denominator of forecasting biodegradation of insecticides {Nakatsugawa and Nelson, 1972}.

The realization of the importance of microsomal oxidase in biodegradation has led Metcalf and co-workers to utilize the synergistic ratio (SR) for assessing biodegradability of insecticides. The SR method {Metcalf, 1967} is based on the near-total inhibition of microsomal oxidases of test organisms by insecticide synergists such as piperonyl butoxide. SR is the ratio of LD_{50} of an insecticide to that of the insecticide synergized with a high dose of a synergist and thus gives a quantitative measure of the significance of oxidative degradation. This method is suited for screening a large number of compounds of the same class after the importance of microsomal oxidation for the class has been established. It has been successfully applied to pyrethroids, carbamates, and DDT analogs. The principle should be applicable to assessing biodegradation by other detoxication enzymes if specific inhibitors are available.

Another method also developed by Metcalf and co-workers to evaluate biodegradability is the model ecosystem {Metcalf et al., 1971}, whereby the biodegradability of a given compound is tested through relatively simple food chains in a small box. Analysis of insecticides and residues in each species and the environment shows the extent of biodegradation and biomagnification. Although these methods are quite useful for examining chemicals for biodegradability, they are not suited for screening a large number of species for biodegradation propensity.

Use of biodegradable insecticides may not automatically guarantee ecological safety. While biodegradation greatly reduces ecological hazards, forecasting of consequences of insecticide use must be accompanied by other considerations. Great variation of enzyme activities is found among animal species. The reported low titer of microsomal oxidases in a snail {Kapoor et al., 1970} may exemplify a case where the use of biodegradable insecticides could fail. Fluctuation of enzyme activities within a species due to physiological changes is well known. Reported instances of biomagnification of parathion (fish and mussels) and diazinon (fish) {Miller et al., 1966} are the evidence that even biodegradable compounds are not foolproof. Furthermore, the threat of ecological damages exists even if the food web transmission of an insecticide does not result in "magnification." Although dramatic mortality or reproductive failure of conspicuous species of birds was found mostly as a result of biomagnification, similar effects can result depending on the sensitivity of the species and the rates of input and elimination without magnification.

Certain biodegradable insecticides owe a part of their degradability to the reactivity of the molecule, as pointed out (Fig. 1) for organophosphates. Such reactive molecules may form covalent bonds with critical biomacromolecules and may have cumulative and long-lasting effects. Although organophosphates are readily degraded and excreted, the macromolecular covalent bonds formed with a minute portion of the dose of the insecticide deserve careful evaluations. No definitive information is available as to the balance sheet of elimination and retention of these chemicals, particularly following repeated exposures to the chemicals. The more transient nature of carbamylation inherent in carbamate insecticides may pose less threat. Relatively fast decarbamylation has been established, however, only with the primary target. Completely reversible inhibitors may be considered safer in the absence of further information. In this context the recent report of reversible (noncarbamylating) inhibition of acetylcholinesterase by aryl N-hydroxy- and N-methoxy-N-methylcarbamates is interesting {Chiu et al., 1973}.

Many substrates for microsomal oxidases are not reactive, but can be converted to reactive forms by microsomal oxidases, i.e., activation of phosphorothionate insecticides, hydroxylation of polycyclic hydrocarbons leading to covalent bond formation with DNA {Gelboin et al., 1972}, dealkylation of 1-aryl-3,-3-dialkyltriazenes, and other activation of precarcinogens {Preussmann et al., 1969; Miller, 1970}.

COMPARATIVE BIOTRANSFORMATION AS A BASIS FOR RISK FORECASTING

The need for comparative data of biotransformation capacity on a wide variety of species is thus evident. As pointed out previously

{Nakatsugawa and Nelson, 1972}, no true comparative studies of biotransformation exist on the enzymatic level and the information is particularly deficient for nonarthropod invertebrates. Only a few species in this category have been studied. The common earthworm, Lumbricus terrestris (Annelida), is equipped with microsomal oxidases as evidenced by metabolism of parathion and epoxidation of aldrin {Nakatsugawa and Nelson, 1972}. The epoxidase is mostly in the intestine and increases from the juvenile to adult stage. Sesamex does not inhibit the enzyme very effectively {Nakatsugawa and Nelson, unpublished results}. Aldrin epoxidase has also been found in the hepatopancreas and alimentary canal of the mussel, Anodonta sp. (Mollusca) and the snail, Lymnaea palustis (Mollusca) {Khan et al., 1972}. In vivo analyses indicate the presence of the epoxidase also in other nonarthropod invertebrates, Hydra littoralis (Coelenterata), a leech (Annelida), and a planaria Dygensia sp. (Platyhelminthes). The oxidative degradation of EPN, p-nitroanisole, aminopyrine, and hexobarbital was not detected in the quahaug, Mercenaria mercenaria (Mollusca) {Carlson, 1972}. This result and the low microsomal oxidase reported for the snail Physa {Kapoor et al., 1970} contrast with the data on epoxidase in the two molluscs just mentioned. Comparison is difficult, however, because of the difference in the substrate, species, age, stage, and the season.

In view of the vast number of animal species, a certain guideline for priority is necessary. Aquatic organisms may deserve special attention since the aquatic ecosystem is more prone to biomagnification, and these animals may possess less microsomal oxidase activity {Hollingworth, 1971}. In considering ecological effects in general, the "major phyla" of the neglected invertebrates as defined in terms of relative biomass {Russel-Hunter, 1969} should probably be considered first.

Survey of microsomal oxidases in a large number of species has been hindered by a variety of obstacles. As is well seen in the studies of insect enzymes {Wilkinson and Brattsten, 1972}, each new species to be examined is different as a rule in the type of endogenous inhibitor of microsomal oxidases, the tissue distribution of enzymes, and the enzyme stability. The small size of many invertebrates often precludes dissection of tissues and necessitates the use of whole animal homogenates, aggravating the problem of endogenous inhibitors.

We have sought a possibility of circumventing these problems in a histological detection of the enzyme. For this purpose one must have a substrate which is converted by microsomal oxidase to a metabolite that is tightly bound to the tissue and is detected in a minute quantity. We have utilized the sulfur-binding phenomenon of parathion activation. Parathion and related compounds are metabolized by microsomal oxidases and produce two sets of metabolites {Nakatsugawa and Dahm, 1967; Neal, 1967}. In one of the reactions paraoxon is produced with the detachment of the sulfur atom which is covalently bound to macromolecules {Nakatsugawa and Dahm, 1967}. The mechanism of the generation of a reactive sulfur atom has been suggested {McBain et al., 1971; Ptashne et al., 1971}. The binding occurs also in vivo {Poore and Neal, 1972}, although the sulfur is eventually excreted as inorganic sulfate {Nakatsugawa et al., 1969}. The reactions are shown in Fig. 4. Our adaptation of this reaction to histological detection of a microsomal oxidase is illustrated in Fig. 5. The sample is freeze-sectioned to 20μ thick, affixed on a slide glass, and covered with ^{35}S-parathion. The uniform thin section will eliminate the problem of penetration of the substrate. After the reaction the substrate and ^{35}S-metabolites other than the bound sulfur are washed away and the specimen is processed for autoradiography. Examples are given in Fig. 6(a-c). Our current effort is in applying this to small organisms

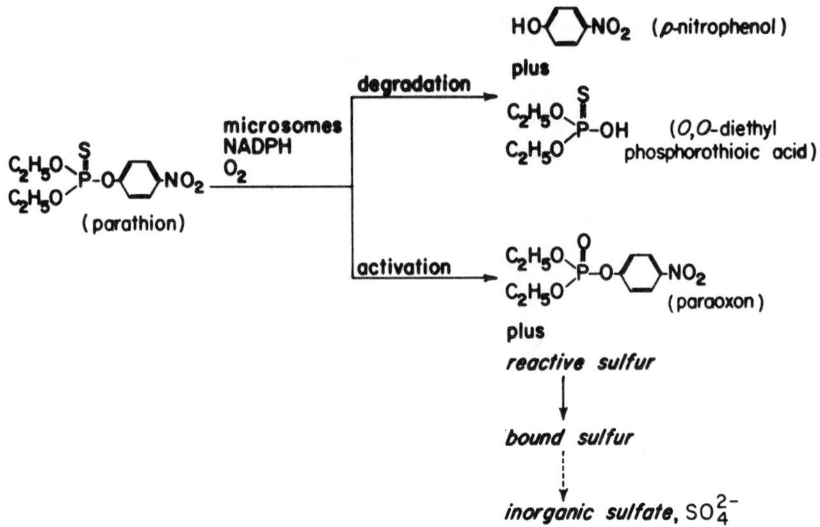

FIG. 4. Microsomal metabolism of parathion and fate of the sulfur atom.

and in refining the technique. If the histological technique is to be used for detection of microsomal oxidases in general, use of more than one substrate would be desirable. Other examples

specimen → freeze in liquid air → section on cryomicrotome → place on slide-glass →

cover with isotonic solution of S-35 parathion —incubation→ remove parathion & fix with formol-alcohol → wash with water, alcohol →

coat with emulsion Kodak NTB 2 —exposure→ develop → stain with Harris hematoxylin eosin Y

FIG. 5. Procedure for histological detection of microsomal oxidases of parathion metabolism.

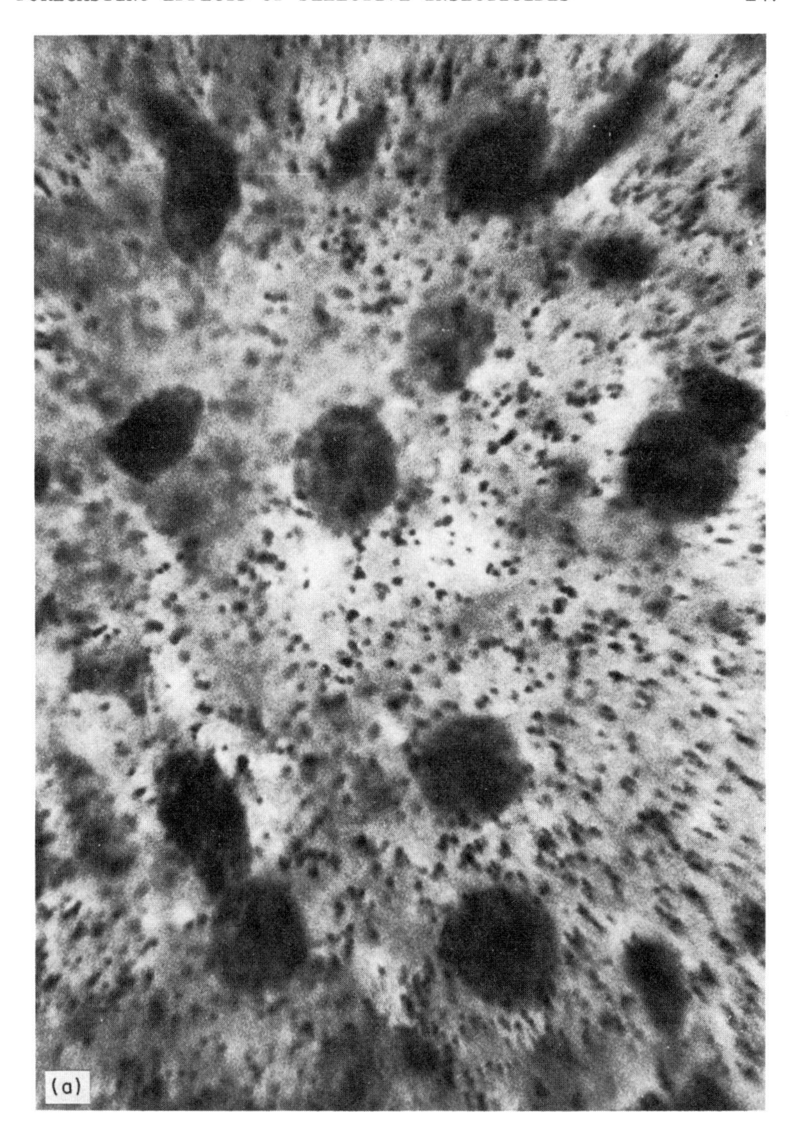

FIG. 6a. Autoradiographic demonstration of microsomal oxidases of parathion metabolism in the male rat liver.

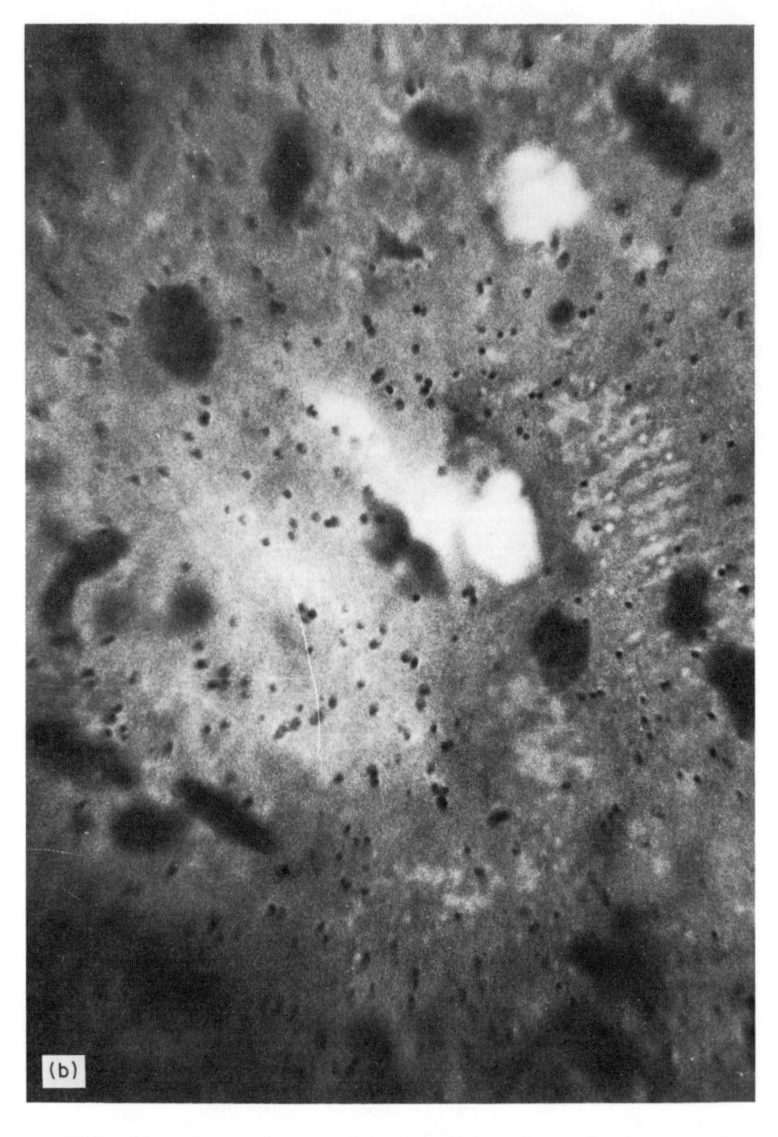

FIG. 6b. Autoradiographic demonstration of microsomal oxidases of parathion metabolism in the house fly fat body.

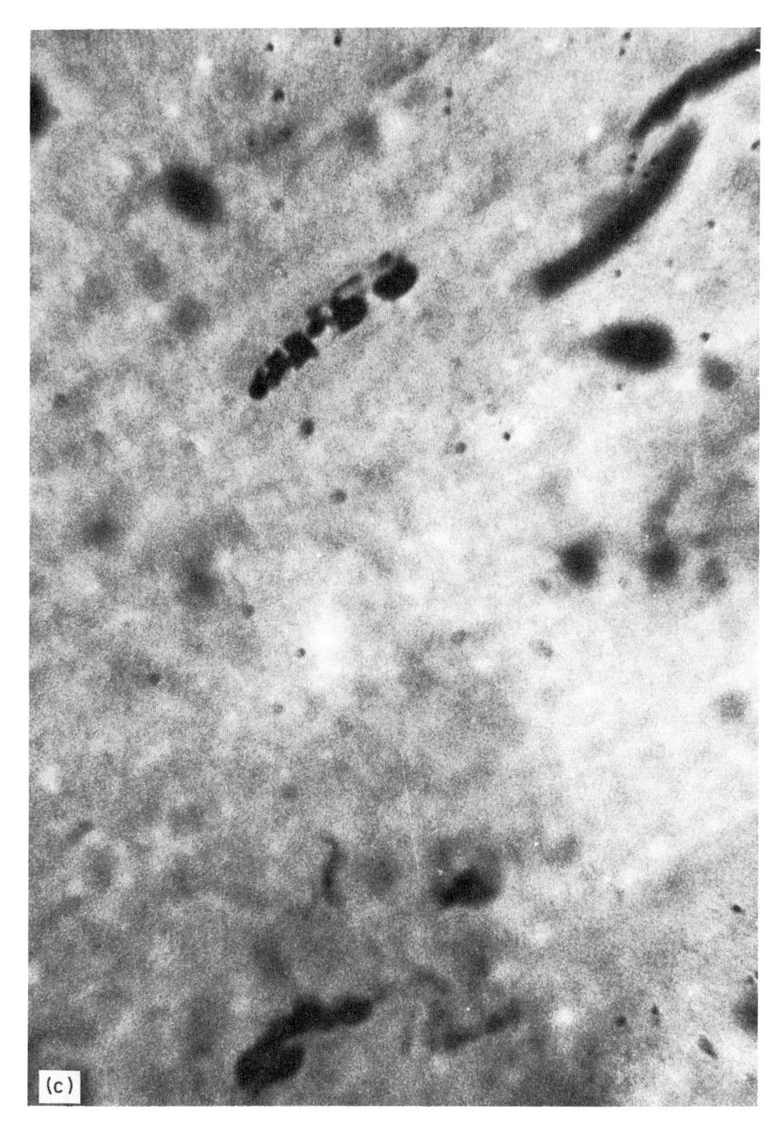

FIG. 6c. Autoradiographic demonstration of microsomal oxidases of parathion metabolism in the house fly flight muscle.

of covalent bond formation due to reactive metabolites of microsomal oxidations are known {Gelboin et al., 1972}.

In summary, comparative studies of biotransformation will provide an important basis for forecasting ecological consequences of selective insecticides. Although collection of such data, particularly on nonarthropod invertebrates, has been hampered by a number of obstacles, information accumulated on vertebrates and insects can help expand the comparative knowledge. Usefulness of such information will depend on the basic studies of the mode of action of insecticides at the levels of primary, secondary, and tertiary biochemical targets. Since some of the secondary and tertiary effects are due to the molecular reactivity which is often reflected in biotransformation, study of biotransformation should reveal not only detoxication mechanisms, but also potential implications of biodegradation reactions in safety. Constant surveillance for unknown side effects is always needed.

ACKNOWLEDGMENTS

*Part of the work described was supported by grants from the Research Foundation of State University of New York 10-7105-B and C. The assistance of H. Appleton, M. Morelli, and R. Smith is gratefully acknowledged.

REFERENCES

Aldridge, W. N., J. M. Barnes, and M. K. Johnson, 1969. Ann. N. Y. Acad. Sci. *160* 314.

Anderson, J. M., 1971. Proc. Roy. Soc. Lond. B. *177*:307.

Bitman, J., H. C. Cecil, S. J. Harris, and G. F. Fries, 1968. Science. *162*:731.

Bitman, J., H. C. Cecil, S. J. Harris, and G. F. Fries, 1969. Nature. *224*:44.

Bois, R. and R. W. Estabrook, 1969. Arch. Biochem. Biophys. *129*:362.

Carlson, G. P., 1972. Comp. Biochem. Physiol. *43B*:295.

Cecil, H. C., J. Bitman, and S. J. Harris, 1971. J. Agric. Food Chem. *19*:61.

Chiu, Y. C., M. A. H. Fahmy, and T. R. Fukuto, 1973. Pest. Biochem. Physiol. *3*:1.

Ernster, L., G. Dallner, and G. F. Azzone, 1963. J. Biol. Chem. *238*:1124.

Eto, M., and H. Ohkawa, 1970. *Biochemical Toxicology of Insecticides*, (R. D. O'Brien and I. Yamamoto, eds.), Academic Press, New York, p. 93.

Fishbein, L., W. G. Flam. and H. L. Falk, 1970. *Chemical Mutagens*, Academic Press, New York, p. 281.

Flockhart, I. J., and J. E. Casida, 1972. Biochem. Pharmacol. *21*:2591.

Gelboin, H. V., N. Kinoshita, and F. Wiebel, 1972. Fed. Proc. *31*:1298.

Gellert, R. J., W. L. Heinrichs, and R. S. Swerdloff, 1972. Endocrinol. *91*:1095.

Heath, R. G., J. W. Spann, and J. F. Kreitzer, 1969. Nature. *224*:47.

Hilgetag, G. and H. Teichmann, 1965. Angew. Chem. *4*:914.

Hilton, B. D., T. A. Bratkowski, T. Yamada, T. Narahashi, and R. D. O'Brien, 1973. Pest. Biochem. Physiol. *3*:14.

Holan, G., 1971. Bull. World Health Org. *44*:355.

Hollingworth, R. M., 1971. Bull. World Health Org. *44*:155.

Horgan, D. J., T. P. Singer, and J. E. Casida, 1968. J. Biol. Chem. *243*:834.

Janicki, R. H., and W. B. Kinter, 1971. Nature (New Biol.) *233*:148

Johnson, M. K., 1969. Brit. Med. Bull. *25*:231.

Kapoor, I. P., R. L. Metcalf, R. F. Nystrom, and G. K. Sangha, 1970. Food Chem. *18*:1145.

Khan, M. A., A. Kamal, R. J. Wolin, and J. Runnels, 1972. Bull. Environ. Contam. Toxicol. *8*:219.

Koch, R. B., 1969. J. Neurochem. *16*:269.

Lofroth, G., 1970. Naturwissenschaft. *57*:393.

Longcore, J. R., F. B. Samson, and T. W. Whittendale, 1971. Bull. Environ. Contam. Toxicol. *6*:485.

Matsumura, F., 1972. *Environmental Toxicology of Pesticides,* (F. Matsumura, G. M. Bousch, and T. Misato, eds), Academic Press, New York, p. 525.

Matsumura, F., T. A. Bratkowski, and K. C. Patil, 1969. Bull. Environ. Contam. Toxicol. *4*:262.

Matsumura, F., and R. D. O'Brien, 1966. J. Agric. Food Chem. *14*:39.

Matsumura, F., and K. C. Patil, 1969. Science. *166*:121.

McBain, J. B., I. Yamaoto, and J. E. Casida, 1971. Life Sciences. *10*:947.

Metcalf, R. L., 1967. Ann. Rev. Entomol. *12*:229.

Metcalf, R. L., T. R. Fukuto, C. F. Wilkinson, M. A. Fahmy, S. A. El-Aziz, and E. R. Metcalf, 1966. J. Agric. Food Chem. *14*:555.

Metcalf, R. L., G. K. Sangha, and I. P. Kappor, 1971. Environ. Sci. Technol. *5*:709.

Miller, J. A., 1970. Cancer Res. *30*:559.

Miller, C. W., B. M. Zuckerman, and A. J. Charig, 1966. Trans. Amer. Fish. Soc. *95*:345.

Nakatsugawa, T. and P. A. Dahm, 1967. Biochem. Pharmacol. *16*: 25.

Nakatsugawa, T., and P. A. Nelson, 1972. *Environmental Toxicology of Pesticides,* (F. Matsumura, G. M. Boush, and T. Misato, eds.), Academic Press, New York, p. 501.

Nakatsugawa, T., N. M. Tolman, and P. A. Dahm, 1969. Biochem. Pharmacol. *18*:1103.

Neal, R. A., 1967. Biochem. J. *103*:183.

O'Brien, R. D., 1967. *Insecticides, Action and Metabolism,* Academic Press, New York, p. 286.

Pimentel, D., 1971. *Ecological Effects of Pesticides on Non-target Species,* Executive Off. of the President. Off. Sci. Technol.

Pocker, Y., W. M. Beng, and V. R. Ainardi, 1971. Science. *174*:1336.

Poore, R. E. and R. A. Neal, 1972. Toxicol. Appl. Pharmacol. *23*:759.

Post, L. C., and W. R. Vincent, 1973. Naturwissenschaften. *60*:431.

Preussman, R., A. von Hodenberg, and H. Hengy, 1969. Biochem. Pharmacol. *18*:1.

Ptashne, K. A., R. M. Wolcott, and R. A. Neal, 1971. J. Pharmacol. Exptl. Therap. *179*:380.

Roger, J-C., D. G. Upshall, and J. E. Casida, 1969. Biochem. Pharmacol. *18*:373.

Rogers, L. J., W. J. Owen, and M. W. Delaney, 1971. Proc. 2nd Internatl. Congress on Photosynth. Stresa, p. 689.

Russel-Hunter, W. D., 1969. *A Biology of Higher Invertebrates*, Macmillan, New York.

Switzer, B., and V. Lewin, 1971. Canad. J. Zool. *49*:69.

Teeter, M. E., M. L. Baginsky, and Y. Hatefi, 1969. Biochem. Biophys. Acta. *172*:331.

Torda, C., and H. Wolff, 1949. J. Pharmacol. Exptl. Therap. *95*:444.

Wang, C. M., and F. Matsumura, 1969. Bull. Environ. Contam. Toxicol. *4*:144.

Wiemeyer, S. N., and R. D. Porter, 1970. Nature. *227*:737.

Wilkinson, C. F., 1971. Bull. World Health Org. *44*:171.

Wilkinson, C. F. and L. B. Brattsten, 1972. Drug Metabolism Rev. *1*:153.

Wilson, W. E., L. Fishbein, and S. T. Clements, 1971. Science. *171*:180.

Yamamoto, I., Y. Soeda, H. Kamimura, and R. Yamamoto, 1968. Agric. Biol. Chem. *32*:1341.

Chapter 8

BASIS FOR SELECTIVITY OF ACARICIDES

Charles O. Knowles
Department of Entomology
University of Missouri
Columbia, Missouri

The order Acarina consists of ticks and mites, and the term acaricide is generally used to designate chemicals active against these organisms {Wharton and Roulston, 1970} (Fig. 1). More specific terms, such as tickicide and miticide, also have been commonly employed. This chapter will focus on the mites, particularly those of agricultural importance. In this instance the term "parasitic" is relegated to those mites which are ecto- and endoparasites of animals; phytophagous is the term used to designate plant-feeding mites (Fig. 1). Parasitic and phyto-

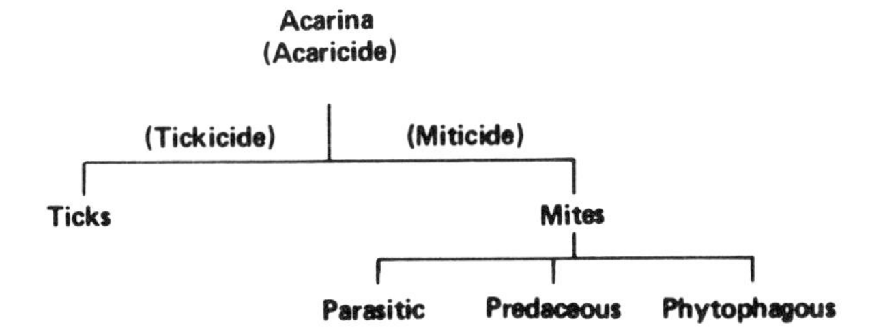

FIG. 1. Terminology used to designate chemicals active against mites and ticks.

phagous mites are primary targets for acaricide chemicals, with the latter being possibly the more significant from an agricultural standpoint. Phytophagous mites are of economic importance on numerous agronomic, horticultural, and ornamental crops. Certain nontarget organisms, including the predaceous mites, are effective in reducing populations of phytophagous mites. Predaceous mites are susceptible to many of the commonly used acaricides; therefore, they also should be included in a consideration of acaricide selectivity. There are other mites, in addition to predaceous, parasitic, and phytophagous species, but they are generally not important in terms of chemical control.

COMPOUNDS POSSESSING ACARICIDAL ACTIVITY

Numerous synthetic organic compounds differing markedly in their chemical configurations possess acaricidal activity {Knowles et al., 1972}. Examples are listed in Fig. 2 according to their respective chemical groups. These include organophosphates (tepp, naled, mevinphos, dicapthon, parathion, phorate, carbophenothion, chlorpyrifos, and azinphosmethyl), carbamates (formetanate, formparanate, carbofuran, aldicarb, SD-17250, and R-17335), chlorinated hydrocarbons (endosulfan and Pentac), nitrophenol derivatives (dinocap, binapacryl, dinobuton, and dinitrocyclohexylphenol), diphenyl aliphatics (dicofol, chlorobenzilate, chloropropylate, and Acarol or bromopropylate), sulfonates (ovex, Genite, and fenson), sulfones (tetradifon), sulfides (tetrasul and chlorbenside), sulfites (Aramite and Omite or propargite), thiodiimides (Milbex), organofluorines (fluenethyl, Nissol, fenazaflor, and R-10044), organotins (Plictran and Du-Ter), formamidines (chlordimeform, C-8520, and H-20013), thioureas (C-9140), phenylhydrazones (Banamite), diaryltriazapentadienes (U-36059), benzohydroximates (Benzomate), arylindones (UC-41305), hydroxybiphenylcarboxanilides (CP-43057), and quinoxalines (oxythioquinox).

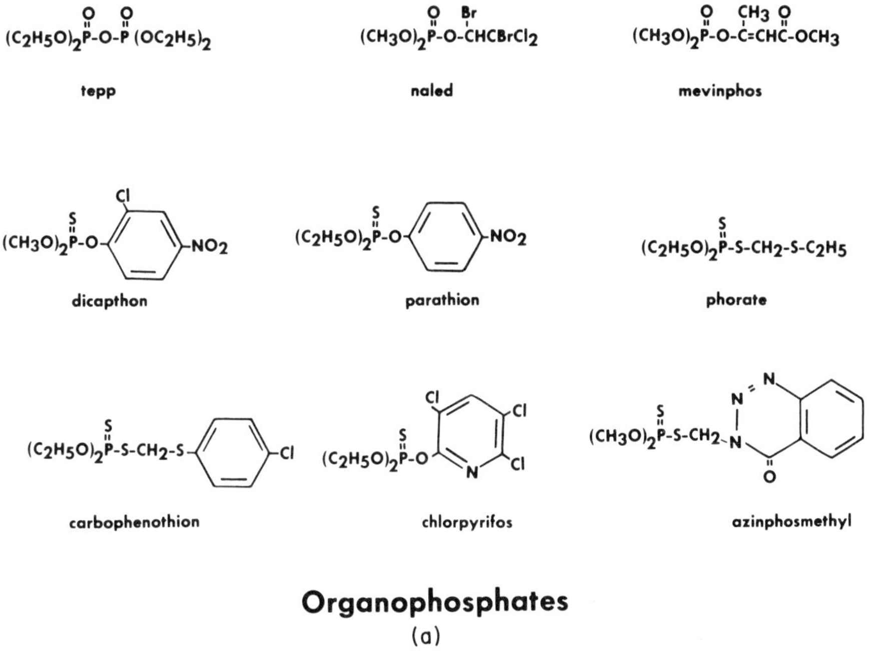

$(C_2H_5O)_2P(O)\text{-}O\text{-}P(O)(OC_2H_5)_2$

tepp

$(CH_3O)_2P(O)\text{-}O\text{-}CH(Br)CBrCl_2$

naled

$(CH_3O)_2P(O)\text{-}O\text{-}C(CH_3)\text{=}CHC(O)\text{-}OCH_3$

mevinphos

$(CH_3O)_2P(S)\text{-}O\text{-}$ Cl, NO_2

dicapthon

$(C_2H_5O)_2P(S)\text{-}O\text{-}$ NO_2

parathion

$(C_2H_5O)_2P(S)\text{-}S\text{-}CH_2\text{-}S\text{-}C_2H_5$

phorate

$(C_2H_5O)_2P(S)\text{-}S\text{-}CH_2\text{-}S\text{-}$ Cl

carbophenothion

$(C_2H_5O)_2P(S)\text{-}O\text{-}$ Cl, Cl, Cl, N

chlorpyrifos

$(CH_3O)_2P(S)\text{-}S\text{-}CH_2\text{-}N$ N N O

azinphosmethyl

Organophosphates

(a)

FIG. 2a. Examples of synthetic organic compounds active as acaricides.

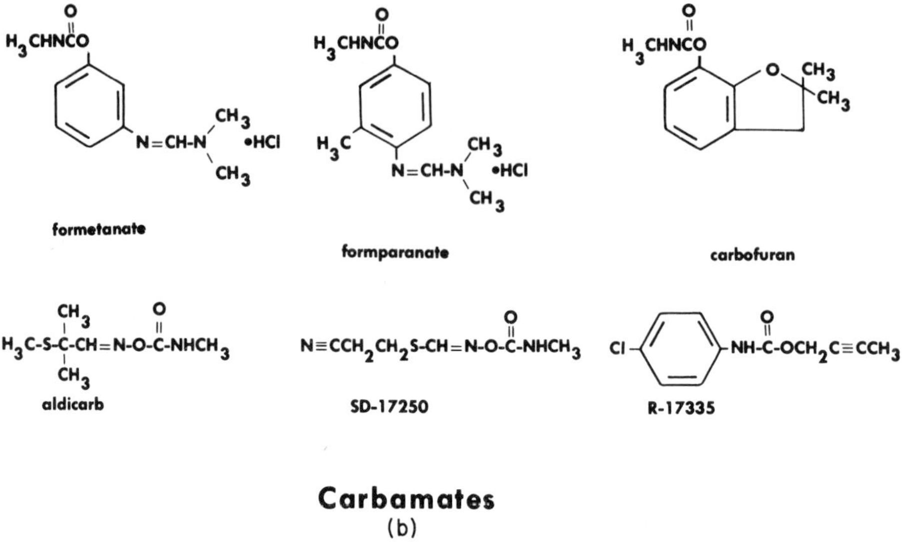

formetanate

formparanate

carbofuran

aldicarb

SD-17250

R-17335

Carbamates

(b)

FIG. 2b. Examples of synthetic organic compounds active as acaricides.

endosulfan

Pentac

Chlorinated Hydrocarbons

(c)

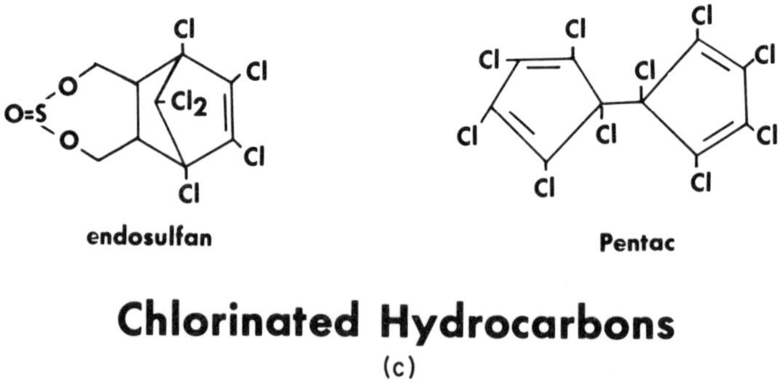

FIG. 2c. Examples of synthetic organic compounds active as acaricides.

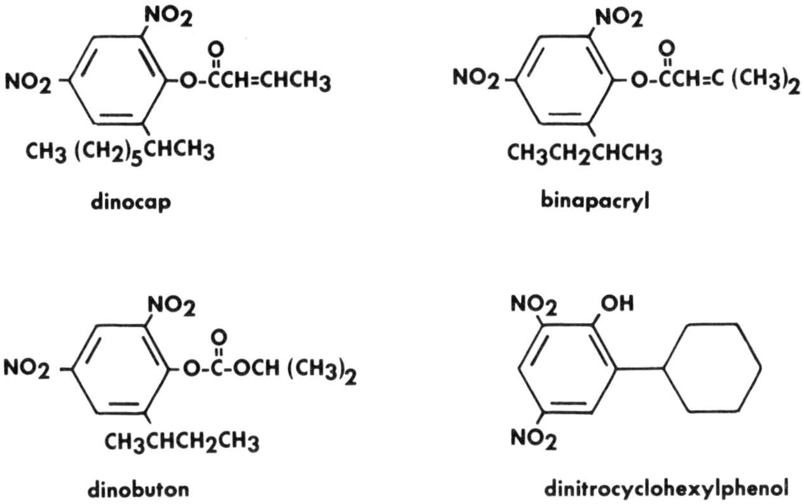

dinocap

binapacryl

dinobuton

dinitrocyclohexylphenol

Nitrophenol Derivatives

(d)

FIG. 2d. Examples of synthetic organic compounds active as acaricides.

dicofol

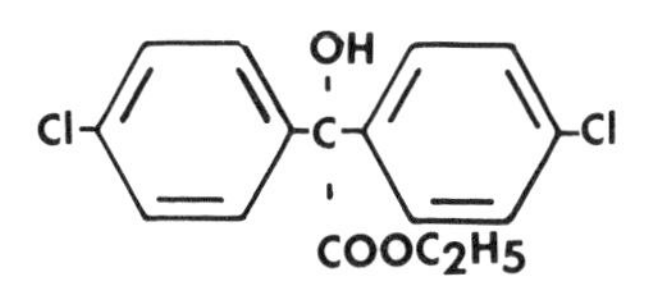

chlorobenzilate

Chloropropylate

Acarol

Diphenyl Aliphatics

(e)

FIG. 2e. Examples of synthetic organic compounds active as acaricides.

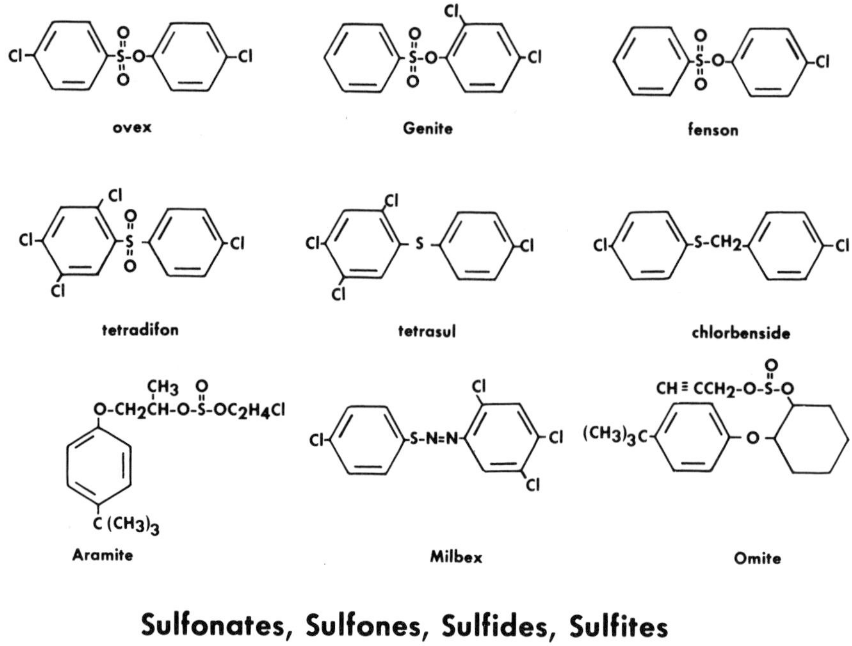

ovex
Genite
fenson
tetradifon
tetrasul
chlorbenside
Aramite
Milbex
Omite
Sulfonates, Sulfones, Sulfides, Sulfites

(f)

FIG. 2f. Examples of synthetic organic compounds active as acaricides.

fluenethyl
Nissol
fenazaflor
R-10044
Organofluorines

(g)

FIG. 2g. Examples of synthetic organic compounds active as acaricides.

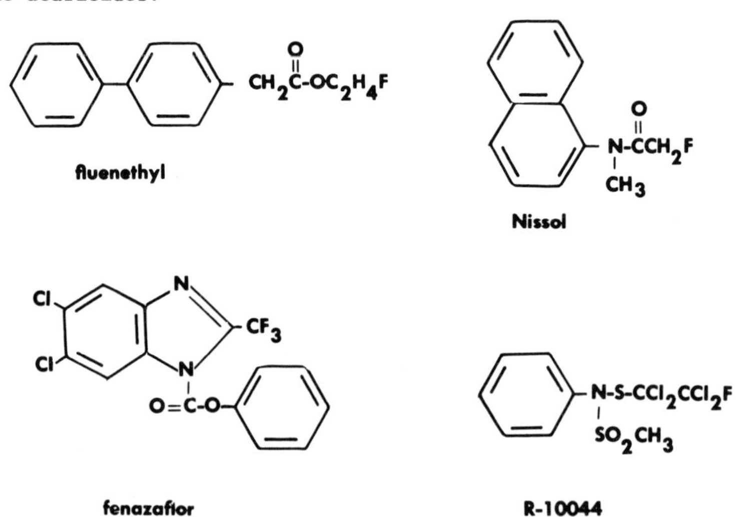

Plictran

Du-Ter

Organotins

(h)

FIG. 2h. Examples of synthetic organic compounds active as acaricides.

chlordimeform

C-8520

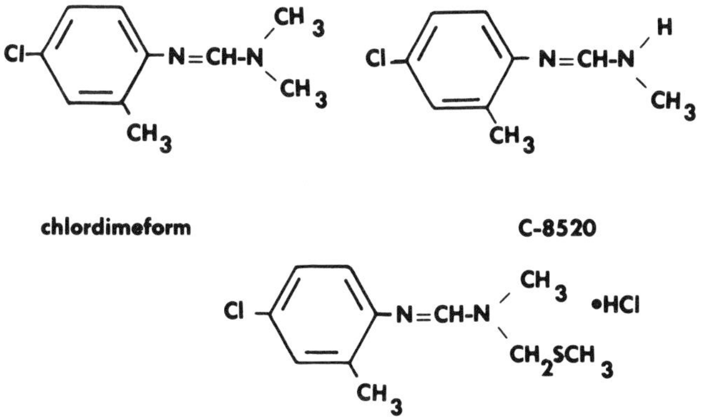

H-20013

Formamidines

(i)

FIG. 2i. Examples of synthetic organic compounds active as acaricides.

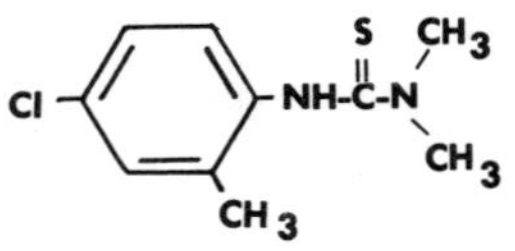

Cl NH-C-N S CH_3 CH_3 CH_3

C-9140

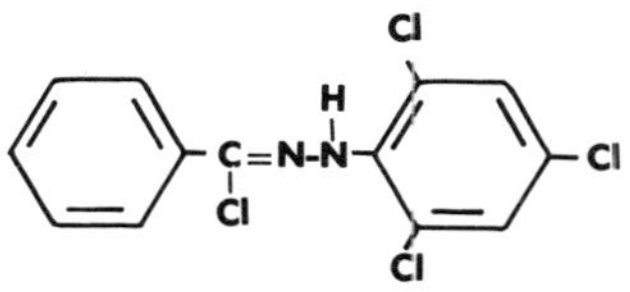

C=N-N H Cl Cl Cl Cl

Banamite

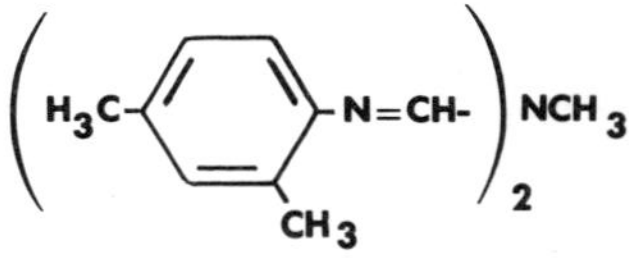

$(H_3C$ N=CH-$)_2$ NCH_3 CH_3

U-36059

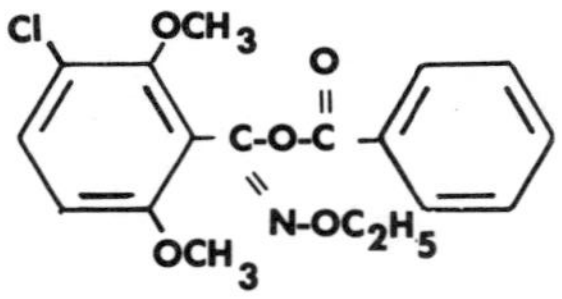

Cl OCH_3 O C-O-C N-OC_2H_5 OCH_3

Benzomate

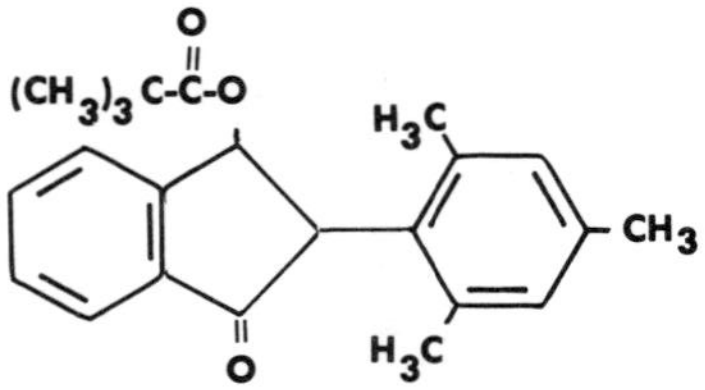

O $(CH_3)_3$ C-C-O H_3C CH_3 H_3C O

UC-41305

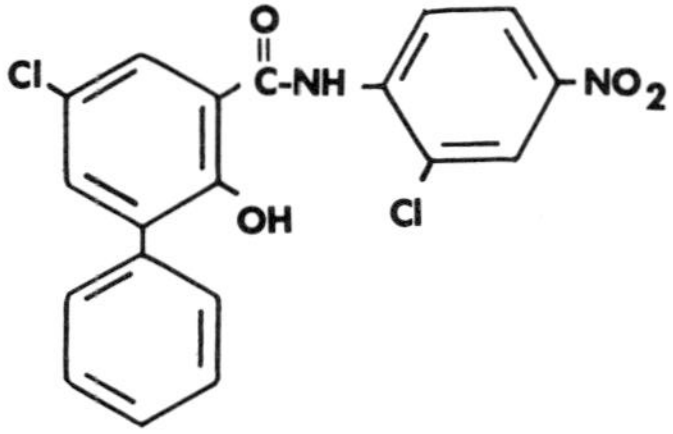

Cl O C-NH NO_2 OH Cl

CP-43057

H_3C N S N S =O

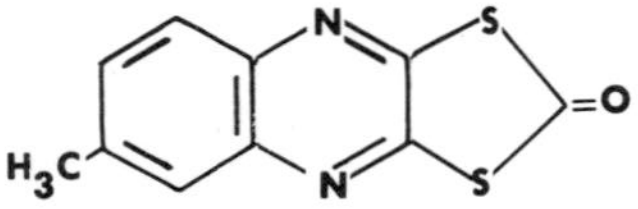

oxythioquinox

Miscellaneous Compounds

(j)

FIG. 2j. Examples of synthetic organic compounds active as acaricides.

SELECTIVITY OF ACARICIDES

From Fig. 2 it is apparent that acaricidal activity occurs among compounds of diverse molecular structures. It is noteworthy that currently we are using commercially more different classes of chemicals as acaricides than as insecticides. Thus, one would expect to find acaricides that run the gamut of the selectivity spectrum, and this appears to be the case. The selectivity classification scheme given in Table 1 will be used as an aid for discussion of the selectivity spectrum of some of these acaricides. There are obvious difficulties associated with this type of classification scheme. However, marked differences in the differential toxicity spectrum of acaricides are to be emphasized, and, to this end, the scheme shown here is acceptable. Compounds listed as Type I generally are nonselective and are highly toxic to numerous mites, insects, and mammals. Type II compounds are moderately selective and are generally toxic to mites and some insects but possess only low to moderate toxicity to mammals. Type III compounds are highly selective and are toxic primarily to mites. One sometimes encounters in the literature the term

Table 1

Selectivity Classification Scheme for Compounds Possessing Acaricidal Activity

Type	Selectivity Level	Toxic to
I	Nonselective	Mites, insects mammals
II	Moderately selective	Mites, insects
III	Highly selective	Mites

"specific acaricide." March {1958} defined a specific acaricide as a compound effective against mites at dosages which are largely ineffective against insects. In the present classification scheme, the Type III and most of the Type II compounds would probably be regarded as specific acaricides.

Table 2 presents the selectivity spectrum of some acaricides. The organophosphates parathion, azinphosmethyl, and mevinphos generally are toxic to mammals, insects, and mites, and, thus, should be classified as Type I or nonselective acaricides. A picture similar to that of the organophosphates is seen with the carbamates formetanate, formparanate, carbofuran, and aldicarb; they are nonselective. The organofluorine fluenethyl is also nonselective. Nissol, fenazaflor, dinobuton, chlordimeform, U-36059, Banamite, and oxythioquinox are toxic to mites but are not so toxic to mammals and most insects. These compounds are categorized as Type II or moderately selective acaricides. Plictran, UC-41305, and the diphenyl aliphatics chlorobenzilate, chloropropylate, and Acarol are also toxic to mites but are generally nontoxic to mammals and insects, except for Plictran which is moderately toxic to mammals. They are classified as Type III or highly selective acaricides.

Up to this point acaricide selectivity has been considered with respect to groups of different organisms, for example mammals, insects, and mites. However, it is well documented that even closely related mite species can differ in their susceptibility to acaricides {Jeppson, 1965}. The main point to be made in this chapter with regard to differential susceptibility of mite species is that it is desirable to have acaricides that can be used in pest management programs under conditions such that harmful mite populations can be reduced to desired levels without appreciably affecting levels of predaceous mites. Problems not only in the use of acaricides but also of insecticides have emerged as a result of killing nontarget predaceous mites.

TABLE 2

Selective Toxicity of Certain Acaricides

Compound	Chemical group	Toxic to[a]			Selectivity	Reference
		Mammal	Insect	Mite	Type[b]	
Parathion	Organophosphate	T	T	T	I	{Negherbon, 1959}.
Azinphosmethyl	Organophosphate	T	T	T	I	{Chemagro Corp, 1961}.
Mevinphos	Organophosphate	T	T	T	I	{Negherbon, 1959}.
Formetanate	Carbamate	T	T	T	I	{Knowles and Ahmad, 1971; Steinhausen, 1968; Wakerly and Weighton, 1969}.
Formparanate	Carbamate	T	T	T	I	{Knowles and Ahmad, 1971; Union Carbide Corp., 1967}.
Carbofuran	Carbamate	T	T	T	I	{Niagara Chem. Div., 1968; Yu et al., 1972}.
Aldicarb	Carbamate	T	T	T	I	{Union Carbide Corp., 1968a}.
Fluenethyl	Organofluorine	T	T	T	I	{Johannsen and Knowles, 1974a; Montecatini Edison Co., 1968}.
Nissol	Organofluorine	MT	MT	T	II	{Johannsen and Knowles, 1972; Nippon Soda Co., Ltd., 1969}.
Fenazaflor	Organofluorine	MT	MT	T	II	{Bowker and Casida, 1969; Fisons Corp., 1969}.
Dinobuton	Nitrophenol Deriv.	MT	MT	T	II	{Bandal and Casida, 1972; Union Carbide Corp., 1968b}.

TABLE 2 (continued)

Compound	Chemical group	Toxic to[a] Mammal	Insect	Mite	Selectivity Type[b]	Reference
Chlordimeform	Formamidine	MT	MT	T	II	{Dittrich, 1966, 1969, 1971; Knowles and Roulston, 1972, 1973; Knowles and Shrivastava, 1973 .
U-36059	Diaryltriaza-pentadiene	MT	MT	T	II	{Knowles and Roulston, 1973; Upjohn Co., 1972}.
Banamite	Phenylhydrazone	MT	MT	T	II	{Knowles and Azia, 1974a; Upjohn Co., 1971}.
Oxythioquinox	Quinoxaline Deriv.	MT	MT	T	II	{Azia and Knowles, 1973b; Sasse, 1960; Wagonner, 1973}.
Plictran	Organotin	MT	LT	T	III	{Ahmad and Knowles, 1972; Gray, 1968; Rock and Yeargan, 1970; Dow Chemical Co., 1973}.
UC-41305	Arylindone	LT	LT	T	III	{Durden and Sousa, 1973; Sousa et al., 1973}.
Chlorobenzilate	Diphenyl Aliphatic	LT	LT	T	III	{Bartsch et al., 1971}.
Chloropropylate	Diphenyl Aliphatic	LT	LT	T	III	{Al-Rubae and Knowles, 1972; Bartsch et al., 1971}.
Acarol	Diphenyl Aliphatic	LT	LT	T	III	{Al-Rubae and Knowles, 1972}.

[a]Based largely on results of acute toxicity. T = generally high toxicity; MT = low to moderate toxicity; LT = extremely low toxicity.

[b]See Table 1, p. 163.

Table 3 shows the toxicity of the specific acaricides (Types II and III) to phytophagous and predaceous mites. These conclusions, based largely on the results of field experiments, show that Plictran, chlorobenzilate, chloropropylate, and Acarol generally possess this kind of selectivity {Bartsch et al., 1971; Gray, 1968; Rock and Yeargan, 1970; Dow Chemical Co., 1973; Westigard et al., 1972}. UC-41305 is a new compound, and published data with regard to its effect on predaceous mites were not found. Although data to allow definitive conclusions were not available in every case, dinobuton, chlordimeform, U-36059, Banamite, and oxythioquinox do not appear to significantly discriminate between phytophagous and predaceous mite species {Upjohn Co., 1971, 1972; Waggoner, 1973}. It should be mentioned that mite-mite selectivity is not restricted to the Type III compounds; some of the other materials cited earlier also can be used in such a manner as to discriminate between mite species {Westigard, 1971; Westigard et al., 1972}.

Stage specificity, another kind of selectivity exhibited by certain acaricides, occurs when an acaricide shows differential toxicity to various stages of the same mite species. For instance, with the two-spotted spider mite, (Tetranychus urticae Koch), chlorobenzilate was equitoxic to the adult and larval stages but was less active against the eggs {Ebeling and Pence, 1954}. There are numerous examples of this particular type of acaricide selectivity {Knowles et al., 1972; March, 1958}.

The last kind of acaricide selectivity is strain specificity, and exists largely because of resistance. Mites have a high reproductive potential, and they may have several generations during a season. Thus, they can be subjected to continuous chemical selection, and resistance can develop rapidly. This resistance phenomenon is the primary reason for so many different classes of acaricides. As resistance develops to

TABLE 3

Toxicity of Specific Acaricides (Types II and III) to Phytophagous and Predaceous Mites

Compound	Chemical group	Toxic to*	
		Phytophagous mites	Predaceous mites
Plictran	Organotin	T	LT
Chlorobenzilate	Diphenyl Aliphatic	T	LT
Chloropropylate	Diphenyl Aliphatic	T	LT
Acarol	Diphenyl Aliphatic	T	LT
UC-41305	Arylindone	T	-
Dinobuton	Nitrophenol deriv.	T	T
Chlordimeform	Formamidine	T	T
U-36059	Diaryltriaza-pentadiene	T	T
Banamite	Phenylhydrazone	T	T
Oxythioquinox	Quinoxaline deriv.	T	T

*Based largely on results of field experiments. See Table 2, p. 165.

currently used acaricides, new compounds presumably with different modes of action are required. Numerous articles on the resistance of acarina to acaricides have been published {Dittrich, 1963; Hansen et al., 1963; Helle, 1965; Jeppson, 1963; Stone, 1972; Wharton and Roulston, 1970; Whitehead, 1965} and others.

BIOCHEMICAL BASIS FOR ACARICIDE SELECTIVITY

The factors that govern the selective toxicity of acaricides are doubtlessly the same as those for other xenobiotics, and they have been enumerated in various publications {O'Brien, 1960, 1961, 1967}. In the case of acaricides, only meager information is available with respect to penetration, mode of action, and metabolism in target and nontarget species. Thus, the biochemical basis for selectivity among acaricides is presently not well understood. However, with other xenobiotics where the mechanism(s) of selective toxicity is at least partially known, differential metabolism seems to be involved in many instances {O'Brien, 1961}. This also appears to be the case with acaricides, notwithstanding the examples of target modification manifested by certain mites and ticks in response to selection with some organophosphates and carbamates {Schunter et al., 1971; Smissaert, 1964; Smissaert et al., 1970; Voss and Matsumura, 1964, 1965; Wharton and Roulston, 1970}. Thus, by taking into account the available data on acaricide metabolism as well as mode of action and structure-activity relationships, it is possible to designate certain chemical groupings in the molecule that could confer an opportunity for selectivity. This point will be illustrated with several examples (see Fig. 3). Only one cleavage site or functional group in each molecule will be designated; others are undoubtedly involved and are obviously important in relation to the total picture of selective toxicity.

Fluenethyl is an organofluorine acaricide of interest because it contains two potential toxaphores, biphenyl and fluorofluenethyl Nissol dinobuton

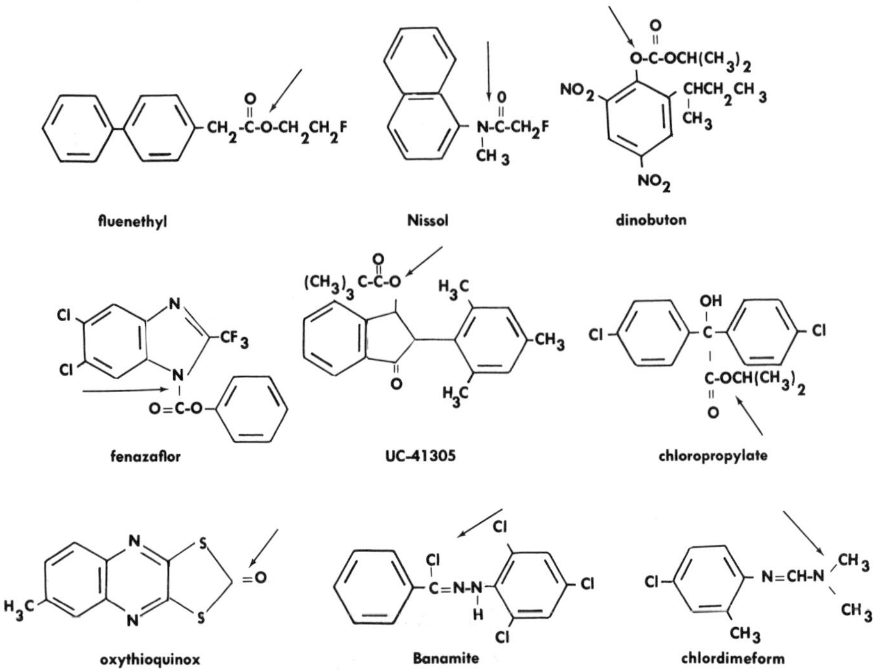

fenazaflor UC-41305 chloropropylate

oxythioquinox Banamite chlordimeform

FIG. 3. Structures of several acaricides showing sites that could confer opportunities for selectivity.

ethanol; it is the only nonselective or Type I compound to be discussed. Fluenethyl metabolism and mode of action studies in mice, house flies, and two-spotted spider mites have shown that cleavage of the ester linkage occurred rapidly in all these species and apparently was requisite for toxicity {Johannsen and Knowles, 1974a, 1974b}. The killing action of this acaricide results from liberation of the monofluoroethanol and its subsequent metabolism to the classical organofluorine aconitase inhibitor {Johannsen and Knowles, 1974a}. Theoretically, the ester linkage in fluenethyl confers an ooportunity for selectivity; however, the actual labile nature of this bond apparently precludes any overt specificity. On the other hand, Nissol is also an organofluorine acaricide, but toxicity studies have indicated some degree of selective action {Hashimoto et al., 1968; Noguchi, Hashimoto, Miyata, 1968}. Animals poisoned with Nissol accumulate citrate indicating aconitase inhibition {Johannsen and Knowles, 1972; Noguchi, Hashimoto, Miyata, 1968}, and *N*-methyl-1-naphthylamine has been identified as a Nissol metabolite, indicating the formation of fluoracetate {Knowles and Shrivastava, 1971; Noguchi, Miyata, Mori, Hashimoto and Kosaka, 1968}. Thus, the mechanism of action is similar to that of fluenethyl, but the amide bond confers selectivity. This situation with Nissol seems similar to that of fluoroacetamide, which is a selective insecticide because of the presence of the amidic bond {Matsumura and O'Brien, 1963}.

Ester and amide bonds are also likely to be important in the selective toxicity of dinobuton and fenazaflor, respectively. Hydrolysis of the ester linkage of dinobuton liberates a dinitrophenol {Bandal and Casida, 1972}, and hydrolysis of the amide bond of fenazaflor liberates a trifluoromethyl-benzimidazole {Bowker and Casida, 1969}. Both of these hydrolysis products uncouple oxidative phosphorylation, and they are much more potent as uncouplers than their respective parent compound {Corbett and Wright, 1970; Ilivicky and Casida, 1969}.

UC-41305 is a highly selective acaricide {Sousa et al., 1973}. From detailed structure-activity studies it was concluded that generation of the parent indandione in situ was necessary for toxicity {Durden and Sousa, 1973}. Thus, it seems likely that the pivaloyloxy linkage is important in the specificity of this acaricide.

Chloropropylate is another highly selective acaricide as are chlorobenzilate and Acarol, two closely related analogs {Al-Rubae and Knowles, 1972; Bartsch et al., 1971}. There is a dearth of information relative to the mode of action of these diphenyl alphatic acaricides. However, activity is dependent on the intact molecule as the halogenated benzilic acid, which is the initial cleavage product in each case, apparently is not acaricidal. Thus, the ester linkage is probably important in determining selectivity.

Oxythioquinox is hydrolyzed to 6-methyl-2,3-quinoxalinedithiol (QDSH) in living organisms {Aziz and Knowles, 1973b; Carlson and DuBois, 1970}; QDSH can inhibit numerous sulfhydryl enzymes {Carlson and DuBois, 1970}. Therefore, the selective action of oxythioquinox may be related to the facility with which it is hydrolyzed to QDSH, although there is some evidence that the parent compound itself also can attack proteins {Aziz and Knowles, 1973b}.

In the case of Banamite, a phenylhydrazone, structure-activity relationship studies have underlined the importance of the benzoyl chlorine atom {Kaugars et al., 1973}. Metabolism studies of Banamite in spider mites revealed that the benzoyl chlorine was replaced by hydroxyl and that a toxic metabolite, 2,4,6-trichlorophenylhydrazine, was subsequently formed {Knowles and Aziz, 1974a, 1974b}. Thus, the benzoyl chlorine could be involved in the specificity of this acaricide.

Chlordimeform is a moderately selective formamidine acaricide. The parent compound is apparently biologically active in some cases {Abo-Khátwa and Hollingworth, 1972; Aziz and Knowles, 1973a; Beeman and Matsumura, 1973}, but indications are that its mono-*N*-demethyl metabolite, demethylchlordimeform (C-8520), may be a potent toxicant {Knowles and Roulston, 1972, 1973; Knowles and Schuntner, 1974}. Piperonyl butoxide abolished chlordimeform toxicity to cattle tick larvae by inhibiting formation of this demethyl metabolite {Knowles and Roulston, 1972, 1973; Knowles and Schuntner, 1974}. Therefore, selectivity of chlordimeform might be related, in part, to removal of an *N*-methyl moiety in the tick.

STRATEGIES FOR THE FUTURE

We frequently hear of the desirability and necessity to develop new pesticides with extremely high target specificity. Some success in this connection has already been achieved with acaricides, as certain of the compounds currently used are selectively toxic to mites. There are numerous mite species of economic importance, and in some cases a single crop can contain simultaneously several different species of phytophagous mites. Thus, the desirability and practicability of developing new acaricides which are selectively toxic to individual mite species seem questionable. A major driving force behind the continued development of new acaricides is the necessity to counter the resistance by mites to the frequently used compounds. In general, these new acaricides should possess the properties desired of insect control agents with respect to environmental compatibility, including minimal side effects on nontarget species. For example, they should be effective against target mites at dosage levels which will not harm predators and pollinating insects such as honeybees. It is noteworthy that chlorobenzilate, one of the diphenyl aliphatic acaricides, has been used successfully to control the acarine disease of bees without affecting the honeybees themselves {Bartsch et al., 1971}. Finally, it is desirable that new acaricides be compatible for use in pest management systems. In this instance a favorable selective toxicity ratio between predaceous mites and phytophagous mites is required.

ACKNOWLEDGMENT

Contribution from the Missouri Agricultural Experiment Station, Columbia. Journal Series No. 6810.

REFERENCES

Abo-Khatwa, N., and R. M. Hollingworth, 1972. Life Sci., part II. *11*:1181.

Ahmad, S., and C. O. Knowles, 1972. Comp. Gen. Pharmacol. *3*: 125.

Al-Rubae, A., and C. O. Knowles, 1972. J. Econ. Entomol. *65*: 1600.

Aziz, S. A., and C. O. Knowles, 1973a. Nature (London). *242*:417.

Aziz, S. A., and C. O. Knowles, 1973b. J. Econ. Entomol. *66*: 1041.

Bandal, S. K., and J. E. Casida, 1972. J. Agr. Food Chem. *20*: 1235.

Bartsch, E., D. Eberle, K. Ramsteiner, A. Tomann, and M. Spindler, 1971. Residue Revs. *39*:1.

Beeman, R. W., and F. Matsumura, 1973. Nature (London). *242*: 273.

Bowker, D. M., and J. E. Casida, 1969. J. Agr. Food Chem. *17*: 956.

Carlson, G. P., and K. P. DuBois, 1970. J. Pharmacol. Exptl. Therap. *173*:60.

Chemagro Corp., 1961. Technical Information on Guthion, Kansas City, Missouri.

Corbett, J. R. and B. J. Wright, 1970. Pestic. Sci. *1*:120.

Dittrich, V., 1963. Adv. in Acarol. *1*:238.

Dittrich, V., 1966. J. Econ. Entomol. *59*:889.

Dittrich, V., 1969. J. Econ. Entomol. *62*:44.

Dittrich, V., and A. Loncarević, 1971. J. Econ. Entomol. *64*: 1225.

Dow Chemical Company, 1973. Plictran Miticide - Technical Information Bulletin, Midland, Michigan.

Durden, J. A., and A. A. Sousa, April, 1973. *Abstracts of Papers*, 165th Meeting ACS, Dallas, Texas.

Ebeling, W., and R. J. Pence, 1954. J. Econ. Entomol. *47*: 789.

Fisons Corp., 1969. Technical Data Sheet for Lovozal Miticide, Wilmington, Massachusetts.

Gray, H. E., 1968. Down to Earth. *23*:3.

Hansen, C. O., J. A. Naegele, and H. E. Everett, 1963. Adv. in Acarol. *1*:257.

Hashimoto, Y., T. Makita, H. Miyata, T. Noguchi, and G. Ohta, 1968. Toxicol. Appl. Pharmacol. *12*:536.

Helle, W., 1965. Adv. in Acarol. *2*:71.

Ilivicky, J., and J. E. Casida, 1969. Biochem. Pharmacol. *18*: 1389.

Jeppson, L. R., 1963. Adv. in Acarol. *1*:276.

Jeppson, L. R., 1965. Adv. in Acarol. *2*:31.

Johannsen, F. R. and C. O. Knowles, 1972. J. Econ. Entomol. *65*:1754.

Johannsen, F. R. and C. O. Knowles, 1974a. Comp. Gen. Pharmacol. *5*:101.

Johannsen, F. R. and C. O. Knowles, 1974b. J. Econ. Entomol. *67*:5.

Kaugars, G., E. C. Gemrich, and V. L. Rizzo, 1973. J. Agr. Food Chem. *21*:647.

Knowles, C. O., and S. Ahmad, 1971. Pest. Biochem. Physiol. *1*:445.

Knowles, C. O., S. Ahmad, and S. P. Shrivastava, 1972. *Pesticide Chemistry, Vol. I* (A. S. Tahori, ed.), Gordon and Breach, London, p. 77.

Knowles, C. O., and S. A. Aziz, 1974a. J. Econ. Entomol. *67*: 574

Knowles, C. O. and S. A. Aziz, 1974b. Bull. Environ. Contam. Toxicol. *12*:158.

Knowles, C. O., and W. J. Roulston, 1972. J. Aust. Entomol. Soc. *11*:349.

Knowles, C. O., and W. J. Roulston, 1973. J. Econ. Entomol. *66*:1245.

Knowles, C. O., and C. A. Schuntner, 1974. J. Aust. Entomol. Soc. *13*:11.

Knowles, C. O., and S. P. Shrivastava, 1971. Unpublished data.

Knowles, C. O., and S. P. Shrivastava, 1973. J. Econ. Entomol. *66*:75.

March, R. B., 1958. Annu. Rev. Entomol. *3*:355.

Matsumura, F., and R. D. O'Brien, 1963. Biochem. Pharmacol. *12*:1201.

Montecatini Edison Co., 1968. Technical Information on New Pesticides--M2060 Winter Ovicide, Milan, Italy.

Negherbon, W. O., 1959. *Handbook of Toxicology*, W. B. Saunders Co., Philadelphia and London, pp. 269, 491, 536, and 580.

Niagara Chemical Division, FMC Corporation, 1968. Technical Report on Furadan, Middleport, New York.

Nippon Soda Company, Ltd., 1969. Technical Data Booklet for Nissol Miticide-Insecticide, Tokyo, Japan.

Noguchi, T., Y. Hashimoto, and H. Miyata, 1968. Toxicol. Appl. Pharmacol. *13*:189.

Noguchi, T., H. Miyata, T. Mori, Y. Hashimoto, and S. Kosaka, 1968. Pharmacometrics. *2*:376.

O'Brien, R. D., 1960. *Toxic Phosphorus Esters*, Academic Press, New York and London, p. 317.

O'Brien, R. D., 1961. Adv. Pest Control Res. *4*:75.

O'Brien, R. D., 1967. *Insecticides--Action and Metabolism*, Academic Press, New York and London, p. 254.

Rock, G. C., and D. R. Yeargan, 1970. Down to Earth. *26*:1.

Sasse, K., 1960. Höfchen-Briefe. *13*:197.

Schuntner, C. A., H. J. Schnitzerling, and W. J. Roulston, 1971. Biochem. Physiol. *1*:424.

Steinhausen, W. R., 1968. Z. Angew. Zool. *55*:107.

Smissaert, H. R., 1964. Science. *143*:129.

Smissaert, H. R., S. Voerman, L. Oostenbrugge, and J. Renooy, 1970. J. Agr. Food Chem. *18*:66.

Sousa, A. A., J. A. Durden, and J. R. Stephen, 1973. J. Econ. Entomol. *66*:584.

Stone, B. F., 1972. Aust. Vet. J. *48*:345.

Union Carbide Corp., 1967. Preliminary Technical Information on UC34096, New York, New York.

Union Carbide Corp, 1968a. Technical Information on Temik, New York, New York.

Union Carbide Corp., 1968b. Technical Information on Dessin, New York, New York.

Upjohn Company, 1971. Technical Information Bulletin for Banamite, Kalamazoon, Michigan.

Upjohn Company, 1972. Technical Information Bulletin for U-36059, Kalamazoo, Michigan.

Voss, G., and F. Matsumura, 1964. Nature (London). *202*:319.

Voss, G., and F. Matsumura, 1965. Canad. J. Biochem. *43*:63.

Waggoner, T. B., 1973. Private communication, Chemagro Corp., Kansas City, Missouri.

Wakerley, S. B., and D. M. Weighton, 1969. Proc. 5th Br. Insectic. Fungic. Conf. 522.

Westigard, P. H., 1971. J. Econ. Entomol. *64*:496.

Westigard, P. H., L. E. Medinger, and D. E. Kellogg, 1972. J. Econ. Entomol. *65*:191.

Wharton, R. H., and W. J. Roulston, 1970. Annu. Rev. Entomol. *15*:381.

Whitehead, G. B., 1965. Adv. in Acarol. *2*:53.

Yu, C.-C., R. L. Metcalf, and G. M. Booth, 1972. J. Agr. Food Chem. *20*:923.

AUTHOR INDEX

A

Abo-Khatwa, N., 172
Ahmad, S., 156, 165, 166, 167
Ainardi, V.R., 140
Albert, A., 114
Aldrich, W. N., 137
Allen, P.M., 36
Al-Rubae, A., 166, 171
Anderson, J.F., 88
Anderson, J.M., 140
Andrews, T.L., 101
Anonymous, 68
Arnold, M.T., 92
Ascher, K.R.S., 80, 82, 83, 84
Aziz, S.A., 166, 172
Azzone, G.F., 140

B

Baginsky, M.L., 140
Baker, G.L., 92
Baker, J.E., 79,83
Bandal, S.K., 165, 171
Barlow, F., 103
Barnes, J.M., 137
Barnes, M.M., 57
Bartlett, B.R., 47
Bartsch, E., 166, 167, 171, 173
Beament, J.M.L., 92
Beck, S.D., 79
Beeman, R.W., 172
Bell, K., 61
Beng, W.M., 140
Bent, K.J., 21, 22
Bergmann, E.D., 86
Beroza, M., 77, 86
Biehn, W.L., 33
Bitman, J., 139, 140
Black, A.L., 101, 103, 104
Blair, E.H., 129
Bodnaryk, R.P., 91
Bois, R., 140
Booth, G.M., 165
Borg, T.K., 79, 83
Bowers, W.S., 86
Bowker, D.M., 165, 171
Bratkowski, T.A., 139
Brattsten, L.B., 145
Brazzel, J.R., 61
Brindley, T.A., 79
British Crop Protection Council, 29
Brown, M.S., 117, 118
Browne, L.B., 78
Bryant, P.J., 86
Burchfield, H.P., 25, 28
Butterworth, J.H., 80, 81

C

Cahill, W.P., 59, 60

Carefoot, G.L., 21
Carlson, G.A., 67
Carlson, G.P., 144, 172
Carter, G.A., 31
Casida, J.E., 94, 127, 137, 138, 140, 145, 165, 171
Casperson, G., 36
Castle, E.N., 67
Cecil, H.C., 139, 140
Chambers, E.E., 88
Chant, D.A., 50, 52
Charig, A.J., 142
Chemagro Corp., 165
Cheng, Hong-Ming, 127
Cheung, L., 86
Chiu, Y.C., 101, 103, 104, 143
Chmurzynska, W., 90
Clemnts, S.T., 139
Clemons, G.P., 37
Clinch, P.G., 100
Cline, R.E., 92
Collins, T.F.X., 29, 30
Corbett, J.R., 171
Cowan, C.B., 49
Crawford, L.G., 97
Crowdy, S.H., 32

D

Dahm, P.A., 145
Dallner, G., 140
Davidase, L.C., 36
Davis, J.W., 49
Davis, K.J., 29
DeBach, P., 54
DeJong, B.J., 93
Dekker, J., 28, 35, 41
Delaney, M.W., 140
Delp, C.J., 38
Dethier, V.G., 78
Dimond, A.E., 33
Dittrich, V., 166, 169
Doane, C.C., 82
Douch, P.G.C., 97
Dow Chemical Co., 166, 167
Driggers, B.F., 56
Druckrey, H. 81
DuBois, K.P., 172
Dunbar, D., 82
Dunn, P.H., 49
Dupras, E.F., 123
Durden, J.A., 166, 171

E

Ebeling, W., 92, 167
Eberle, D., 166, 167, 171, 173
Edelman, N.M., 90
Edgington, L.V., 26, 34
El-Aziz, S.A., 137
Elliott, M., 96
Endo, A., 89
Ernster, L., 140
Estabrook, R. W.
Estesen, B.J., 59, 60
Eto, M., 97, 99, 138
Etzel, L.K., 49
Evans, E., 32, 33, 36, 43
Everett, H.E., 169

F

Fahmy, M.A.H., 101, 103, 104, 137, 143
Falcon, L.A., 49
Falk, H.L., 138
Ferkovich, S.M., 79, 83
Ferris, C.A., 49
Fishbein, L., 138, 139
Fisons Corp., 165
Fitzpatrick, T.B., 90
Flam, W. F., 138
Flint, W.P., 49
Flockhart, I.J., 137
Flower, L.S., 103
Fraenkel, G., 78
Fraser, J., 100, 103
Fries, G.F., 139, 140
Frost, K.R., 59, 60
Fukuto, T.R., 49, 97, 99, 101, 103, 104, 137, 143
Fullerton, D.G., 60

G

Gaborjanyi, R., 83
Gaines, T.B., 69
Garcia, Fernandiz, F., 36
Gasser, R., 48
Gaston, L.K., 77
Gelboin, H.V., 143, 150
Gellert, R.J., 139
Gemrich, E.C., 172
Gerhardt, P.D., 59, 60
Gerschenfeld, H.M., 86
Getzin, L.W., 49
Gijswijt, M.J., 93
Gilbert, B.L., 79
Gill, J.S., 82
Gorz, H.J., 80
Gottlieb, D., 36
Graham, S.L., 29, 30
Gray, H.E., 166, 167
Green, M.B., 5
Greenslade, R.M., 74
Greenwood, D., 103
Grose, J.E.H., 103

H

Hadaway, A.B., 103
Hagen, K.S., 47, 49, 51, 53
Hall, I.M., 49
Hammerschlag, R.S., 36
Hansberry, R., 68
Hansen, C.O., 169
Hansen, W.H., 29
Harris, S.J., 139, 140
Harrison, I.R., 103
Hartley, G.S., 74
Hashimoto, Y., 171
Haskell, P.T., 84
Haskins, F.A., 80
Hatefi, Y., 140
Headley, J.C., 67
Heath, R.G., 139
Heinrichs, W.L., 139
Helle, W., 169
Hengy, H., 143
Hennessy, D.J., 103
Henrick, C.A., 86
Hilgetag, G., 138
Hilton, B.D., 139
Hock, W.K., 35
Holan, G., 137
Hollingworth, R.M., 95, 97, 99, 135, 144, 172
Hook, G.E.R., 97
Hooker, A.L., 22
Horgan, D.J., 140
Hori, M., 88
Hoyt, S.C., 49, 56, 57, 59

I

Ilivicky, J., 171
Ishaaya, I., 83, 84, 94
Ivankovic, S., 81

J

Jackson, L.L., 92
Jacobson, M., 77
Jain, M.K., 80
Janicki, R.H., 139
Jefferson, R.N., 77
Jeppson, L.R., 164, 169
Jermy, T., 78, 82, 83
Johannsen, F.R., 165, 170, 171
Johnson, E.R., 93
Johnson, M.K., 137, 138
Johnson, O., 68, 69, 71
Jumar, A., 82

K

Kaars Sijpesteijn, A., 26, 32, 36
Kakiki, K., 88
Kamal, A., 144
Kamimura, H., 140
Kaplanis, J.N., 86
Kapoor, I.P., 142, 144
Kaugars, G., 172
Kellogg, D.E., 167
Khan, M.A., 44
Kimmel, E.C., 86

Kinoshita, N., 143, 150
Kinter, W.B., 139
Klopping, H.I., 38
Klun, J.A., 79
Kneese, A.V., 67
Knowles, C.O., 156, 165, 166, 167, 170, 171, 172
Koch, R.B., 139
Koeppe, J.K., 92
Kohn, G.K., 115, 116, 117, 118
Kosaka, S., 171
Kreitzer, J.F., 139
Krishnakumaran, A., 86

L

Lapidus, J.B., 79
Large, E.C., 21, 22, 23
Lavie, D., 80
LeClerg, E.L., 21
Leigh, T.F., 49
Lerner, A.B., 90
Lewallen, L.L., 91
Lewin, V., 139
Lewis, C.T., 82
Lewis, D.K., 103
Lingren, P.D., 49
Locke, M., 92
Lofroth, G., 138
Loncarevic, A., 166
Long, J.W., 25, 27, 28
Longcore, J.R., 139
Look, M., 101
Lord, F.T., 49
Loschiavo, S.R., 81
Love, F., 88
Lukens, R.J., 26, 27, 31, 34, 35
Lyle, C., 53
Lyn, H., 36
Lyon, R.L., 101

M

McBain, J.B., 145
McCallan, S.E.A., 25, 28
McFarlane, J.E., 90
McMullen, R.D., 91
MacPhee, A.W., 49
Maddrell, S.H.P., 87
Madsen, H.F., 57
Makita, T., 171
Manglitz, G.R., 80
March, R.B., 101, 103, 164, 167
Marsh, R.W., 39
Martin, D.F., 61
Marx, J.L., 77
Mathre, D.E., 36, 37
Matolcsy, G., 82, 83
Matsumura, F., 139, 169, 171, 172
Matteson, J., 82
Medinger, L.E., 167
Mehrotra, K.N., 97, 99
Meltzer, J., 93
Menn, J.J., 86
Metcalf, C.L., 49
Metcalf, E.R., 137
Metcalf, R.A., 96, 100
Metcalf, R.L., 29, 48, 68, 69, 76, 96, 99, 100, 103, 137, 141, 142, 144, 165
Miller, C.W., 142
Miller, J.A., 143
Miller, L.P., 25, 28
Mills, R.R., 92
Misato, T., 88, 89
Miskus, R.P., 101
Miyamoto, J., 97
Miyata, H., 171
Montecatini Edison Co., 165
Moore, J.E., 115, 116, 118
Mordue, A.J., 84
Morgan, E.D., 80, 81
Mori, T., 171
Moscowitz, J., 82
Mulder, R., 94
Munakata, K., 81
Minniappan, R., 81
Munoz, E., 36
Myers, R.O., 165

N

Naegele, J.A., 169

Nakajima, M., 77
Nakatsugawa, T., 141, 144, 145
Narahashi, T., 139
National Academy of Science, 81
Neal, R.A., 145
Negherbon, W.O., 165
Nelson, P.A., 141, 144
Newson, L.D., 49, 53
Niagara Chemical Division, 165
Nippon Soda Company, 165
Noguchi, T., 171
Norris, D.M., 79, 83
Nystrom, R.F., 142, 144

O

O'Brien, R.D., 76, 86, 96, 136, 139, 169, 171
Ogita, Z., 90
Ohkawa, H., 138
Ohta, G., 171
Oostenbrugge, L., 169
Ospenson, J.N., 115, 116, 118
Owen, W.J., 140

P

Patil, K.C., 139
Pence, R.J., 167
Percy, G.R., 81
Perry, C.H., 29
Persing, C.O., 68
Peters, D.C., 92
Peterson, G.D., 49
Phokela, A., 97, 99
Pickett, A.D., 49, 56, 59
Pimentel, D., 29, 137
Pitman, R.M., 86, 87
Pocker, Y., 140
Poore, R.E., 145
Posnova, A.M., 90
Post, L.C., 93, 136
Preussman, R., 81, 143
Ptashne, K.A., 145

Q

Quraishi, M.S., 91

R

Ragsdale, N.N., 36
Rainwater, C.F., 82
Ramsteiner, K., 166, 167, 171, 173
Ready, R.C., 100
Regnier, F.E., 92
Renooy, J., 169
Reynolds, H.T., 49
Ridgway, R.L., 49
Ripper, W.E., 53, 74
Ritter, G., 36
Rizzo, V.L., 172
Robbins, W.E., 86
Rock, G.C., 166, 167
Roger, J.C., 137, 138
Rogers, L.J., 140
Roulston, W.J., 155, 166, 169, 171, 172
Rozental, J.M., 79, 83
Runnels, J., 144
Ruscoe, C.N.E., 81
Russel-Hunter, W.D., 144

S

Sacher, R.M., 94
Sakata, M., 97, 99
Samson, F.B., 139
Sanford, K.H., 49
Sangha, G.K., 142, 144
Saringer, G., 83
Sasayama, T., 97, 99
Sasse, K., 166
Schaefer, C.H., 94, 101, 104, 123
Schneiderman, H.A., 86
Schnitzerling, H.J., 169
Schooley, J.M., 34
Schoonhoven, L.M. 84
Schrader, G., 96

Schuntner, C.A., 169, 172
Sehnal, F., 86
Shea, K.P., 29, 30
Sherald, J.L., 36
Shorey, H.H., 77
Shrivastava, S.P., 156, 166, 167, 171
Shuval, G., 84
Siddall, J.B., 86
Siegel, M.R., 25, 27, 28, 34
Silverstein, R.M., 77
Singer, T.P., 140
Sisler, H.D., 25, 27, 28, 35, 36, 37
Skaff, V., 91
Slama, K., 86
Slossner, J.E., 61
Smissaert, H.R., 169
Smissman, E.E., 79
Smith, C.N., 78
Smith, J.N., 97
Smith, R.F., 47, 48, 49, 51, 53, 54
Soeda, Y., 140
Solel, Z., 26, 34
Somers, E., 25
Sousa, A.A., 166, 171
Spann, J.W., 139
Specht, H.B., 56
Sphan-Gabrielith, S.R., 80
Spielman, A., 91
Spindler, M., 166, 167, 171, 173
Sprott, E.R., 21
Staal, G.B., 86
Steinhausen, W.R., 165
Stephen, J.R., 166, 171
Stern, V.M., 47, 49, 50, 51, 53
Sternburg, J., 87
Stevick, L., 115, 116, 118
Stinner, R.E., 49
Stone, B.F., 169
Stringer, A., 38
Stronberg, L.K., 49
Sun, Y.P., 93
Suzuki, S., 88
Svoboda, J.A., 86
Swerdloff, R.S., 139
Switzer, B., 139

T

Taft, H.M., 82
Teeter, M.E., 140
Teichmann, H., 138
Thomas, H.Z., 56
Thompson, M.J., 86
Thomson-Hayward Chemical Co., 93
Thorsteinson, J.J., 78
Tillman, R.W., 25, 27, 28
Tipton, C.L., 79
Tomann, A., 166, 167, 171, 173
Tolman, N.M., 145
Torda, C., 139
Turner, C.R., 103
Turpin, F.T., 92
Twinn, D.C., 86

U

Ulrich, J.T., 36, 37
Union Carbide Corp., 165
Unterstenhofer, G., 73
Upjohn Co., 166, 167
Upshall, D.G., 137, 138
U.S. Dept. of Agriculture, 42
U.S. Dept. of HEW, 1, 3, 8

V

van Daalen, J.J., 94
van de Bosch, R., 47, 49, 50, 51, 53
Vardanis, A., 97
Vasquez, D., 36
Vincent, P.G., 26, 28
Vincent, W.R., 93, 136
Voerman, S., 169
Von Hodenberg, A., 81, 143
Voss, G., 169

W

Waggoner, T.B., 166, 167
Wain, R.L., 31
Wakerley, S.B., 165
Wallis, R.C., 90
Walker, W.F., 86
Wang, C.M., 139
Ware, G.W., 59, 60
Watson, T.F., 55, 60, 61
Watt, K.E.F., 52
Weed Science Society, America, 4
Weibel, F., 143, 150
Weiden, M.H.J., 103
Weighton, D.M., 165
Weinberg, A.W., 113
Wellinga, K., 94
Wellman, R.H., 25, 42
Wells, W.H., 103
Wescott, E.W., 37
Westigard, P.H., 49, 56, 57, 59, 167
Wharton, R.H., 155, 169
Whitcomb, W.H., 61
Whitehead, G.B., 169
Whittendale, T.W., 139
Wiebel, F., 143, 147
Wiemeyer, S.N., 139
Wilder, W.H., 94, 101, 104, 123
Wilkinson, C.F., 137, 145
Wilson, W.E., 139
Winston, P.W., 92
Winteringham, F.P.W., 76, 86
Witt, J.M., 60
Wojtczak, L., 90
Wolcott, R.M., 145
Wolff, H., 139
Wolin, R.J., 144
Wood, D.L., 77
Woodcock, D., 33, 38
Wright, B.J., 171
Wright, D.P., Jr., 79, 81, 82, 84
Wright, M.A., 38

Y

Yamada, T., 139
Yamamoto, I., 140, 145
Yamamoto, R., 140
Yeargan, D.R., 166, 167
Yu., C.-C., 165

Z

Zuckerman, B.M., 142

SUBJECT INDEX

A

Aatrex, *see* Atrazine
Abate, 122
AC-24,055 {4'-(3,3-dimethyl-1-triazeno)acetanilide}:
  insect antifeedant, 84
  mechanism of action, 84
  structure of, 80
Acaricides:
  biochemical basis for selectivity of, 169-172
  definition of, 155
  practicalities for selective use, 172,173
  role in fruit pest management, 57
  selectivity among acarina, 163-169
  selectivity classification, 163
  selectivity vs. organophosphates and carbamates, 164
  "specific" acaricide, 164
  stage (in life) specificity, 167
  strain (genetic) specificity, 167
  resistence to, 173
Acarol:
  acaricidal activity, 156,164, 166,167,168,171
  structure of, 159
  toxicological selectivity, 166,167
Acetylcholine esterase comparative inhibition by selective insecticides, 97
Aedes nigromaculis, 123,124, 126
Alachlor:
  crop uses, 4
  weeds controlled, 4
Aldicarb:
  acaricidal activity, 156, 164,165
  insecticidal selectivity, 100
  structure of, 159
  toxicological selectivity, 165
Aldrin:
  metabolism by non-arthropod invertebrates, 144
  toxic properties of, 69
Alkyl phenyl carbamates:
  structure activity relations, 115,120
  toxicity index of, 116
Alternaria tenuis, 25
Anodonta spp., 104
Anopheles stephensi, 103
Antifeedants estimated role in selective insect control, 84,85

Aramite:
  acaricidal activity, 156
  structure of, 160
Arylindone acaricides, *see* UC-41305
Ascomycetes systemic fungicidal selectivity, 32,33
Atrazine weed control and crop uses, 4
Azdirachtin, neem tree principal
  insect antifeedant, 80,81,82
  mechanism of antifeedant action, 84
Azinphosmethyl:
  acaricidal activity, 156
  precision application, 61
  in codling moth control, problems with predators of mice, 57,58
  structure of, 157
  toxicological selectivity, 165
Azodrin, 69

B

Banamite:
  acaricidal activity, 156,164, 166,167,168
  mechanism of selective toxicological action, 172
  structure of, 162,170
  toxicological selectivity, 166
Basidiomycetes systemic fungicidal selectivity, 32,33
Baygon as bollweevil antifeedant, 82
Baytex *see* Fenthion
Bean aphid, 103
Beet Army Worms outbreak after carbaryl use on cotton, 55
Beneficial insects, 12
Benomyl:
  mechanism of fungicidal action, 36
  mammalian toxicity, 38
  phytotoxic effect, 33
  toxicity to earthworms and miticide, 38
Benzimidazoles *also see* benomyl, 32,33
Benzomate:
  acaricidal activity, 156
  structure of, 162
Benzoylphenylureas:
  as insecticides, 89
  effects on insect molting, 93
  structure-activity relations, 94
Binapacryl:
  acaricidal activity, 156
  structure of, 159
Biodegradation of insecticides:
  as factor in selective toxicity, 141-146,150
  in model ecosystems, 142
  reactivity of products, 143
  species comparisons as basis for risk taking, 143-146, 150
Biomagnification of toxicants, 141-143
Biphenylcarboxanilide acaricides *see* CP-43057
Birds:
  effects on reproduction of, 139
  egg shell thinning, 139
Blasticidin:
  mammalian toxicity, 38
  pathogen resistance mechanism, 35
Blatella germanica, 121
Blowfly larvae, 102
Bordeaux mixture, 22, 23
Brestan:
  insect antifeedant, 81,82
  structure of, 80
Bromopropylate *see* acarol
Bursicon, 88
Butacarb, 100
Bux, insecticide, 69,116,123, 124,126,127

C

C-8520 (Formamidine acaricide):
  acaricidal activity, 156
  structure of, 161
C-9140 (Formamidine acaricide):
  acaricidal activity, 156
  structure of, 162
Captafol, 26
Captan:
  different uptake by pathogens, 37
  fish toxicity, 29
  fungicidal activity, 24,25,26, 29,35,37
  mutagenic, teratogenic, 29
  mammalian toxicity, 29
  soil use, 30
Carbamate insecticides:
  biscarbamayl sulfide analogues, 101,102
  derivatization to improve selective toxicity, 101-104
  derivatives; possibilities in usage, 105
  N-phosphoro analogues, 101, 102
  secondary hazards via carbamylations, 143
Carbaryl, 69,100
Carbofuran:
  acaricidal activity, 156,164, 165
  N-arylsulfenyl derivative as insecticide, 101
  structure of, 158
  toxicological selectivity, 165
Carbophenothion:
  acaricidal activity, 156
  structure of, 157
Carboxin:
  fungicidal selectivity, 33
  mechanism of fungicidal action, 36
  mammalian toxicity, 38
Carcinogenicity as a tertiary insecticide action, 137
Cattle tick, 172
Cell wall formation as affected by inhibited chitin synthesis, 37
Chitin synthesis:
  as insecticide target, 87, 88
  fungicidal inhibition, 36,37
  insect, inhibition of in molting, 93
Chitinase, insect stimulation by benzoylphenylureas, 93
Chlorbenside:
  acaricidal activity, 156
  structure of, 160
Chlorbenzilate:
  acaricidal activity, 156,164, 166,167,168,171
  safety to pollinators, 173
  structure of, 159
  toxicological selectivity, 166
Chlordane, 69
Chlordimeform:
  acaricidal activity, 156, 164,166,167,168
  as antifeedant, 82
  mechanism of selective toxicity, 172
  structure of, 170
  toxicological selectivity, 166
Chlorinated insecticides side effects, 138
Chlorpyrifos:
  acaricidal activity, 156
  structure of, 157
Chloroneb:
  fungicidal selectivity, 32, 33
  mammalian toxicity, 38
  mechanism of fungicidal action, 35
Chloronil, 26
Chloropicrin, 30
Chloropropylate:
  acaricidal activity, 156, 164,166,167,168
  mechanism of selective toxicity, 171

structure of, 159,170
toxicological selectivity, 166,168
Chlorothalonil:
fish toxicity, 29
fungicidal activity, 24,25,26
mammalian toxicity, 29
mechanism of action, 27
reactivity with proteins, 28
Colletotrichum coffeanum (coffee bean disease), 21
Comparative toxicology need for in design of selective insecticides, 95
Copper (fungicides):
fungicidal activity, 25
mammalian toxicity, 29
Cotton leaf perforator outbreak after carbaryl use on cotton, 54,55
Cotton loopers outbreak after carbaryl use on cotton, 55
Coumarins and coumarin precursors insect antifeedants, 79
CP-43057 (Formamidine acaricide):
acaricidal activity, 156
structure of, 162
Crabgrass (*Digitaria* sp.), 5
Culex sp., 102,103
Culex tarsalis, 123,124,125, 126
Culex pipieus q., 123,124,125
Cycloheximide:
different uptakes by pathogens, 37
fungicidal activity, 36
mammalian toxicity, 38
Cyperus rotundus L. *see* nut sedge

D

2,4-D in nut sedge control, 5
Dasanit
D-D, 30
DDT:
mechanism of toxic action, 139
secondary toxic effects, 139, 140
Delayed neurotoxicity secondary effect of insecticides, 137
2,6-di-*t*-butyl-4 (α,α-dimethylbenzyl)phenol selective mosquito larvicide, 94
Diazinon:
biomagnification in non-target organism, 142
Toxicological properties, 69
Diazoben soil use, 30
Dibrom, 126
Digitaria spp., 5
Dicapthon:
acaricidal activity, 156
structure of, 157
Dichlone:
fungicidal activity, 24,26
mammalian toxicity, 29
Dicloran, 24
Dichlorovos mutagenic action, 138
Dicofol
acaricidal activity, 156
structure of, 159
2,4-dihydroxy-7-methoxy-1,4-(benzoxazin-3-one) insect antifeedant, 79
3-(3,4-dihydroxyphenyl)-2-hydrozino-2-methylpropionic acid DOPA decarboxylase inhibitor as insecticide, 91
DIMBOA *see* 2,4-dihydroxy-7-methoxy-1,4-(benzoxazin-3-one)
Dimethyl dithiocarbamate, 26
4'-(3,3-dimethyl-1-triazeno) acetanilide *see* AC-24,055
Dinitrocyclohexylphenol:
acaricidal activity, 156
structure of, 159
Dinobuton:
acaricidal activity, 156,164, 165,167,168
mechanism of selective toxicity, 171

structure of, 159,170
toxicological selectivity, 165,168
Dinocap:
acaricidal activity, 156
fungicidal activity, 24
structure of, 159
Disulfaton, 69
Dithiocarbamates *see also* maneb, zineb, ziram:
cuticular insecticides, phenoloxidase inhibition by, 90
fungicidal activity, 26
soil fungicide use, 30
Diuron *see* Karmex
Dodine, 24,25
DOPA decarboxylase role in cuticle tanning, 91
Drosophila sp., 91
Drozoxolon, 24
Dursban, 69,126
Du-Ter:
acaricidal activity, 156
insect antifeedant, 81,82
structure of, 80,161
Dygensia sp., 144
Dyrene:
fungicidal activity, 24,26
mammalian, fish toxicity, 29

E

Earthworm (*Lumbricus terrestris*) toxicant metabolism by, 144
Ecdysiotropin insect hormone, 88
Ecdysones cuticle tanning, 91
Economics:
externalities in pesticide use, 67
of pesticide development, 6, 85,129,130,131,132
Egyptian cotton leafworms, 104
Emergence of minor insect species to pest status, 53
Endosulfan:
acaricidal activity, 156
structure of, 158
Entomologists trained for pest management, need for, 62
Environment effects of insecticides, 12
Environmental pollution narrow spectrum vs. broad spectrum herbicides, 7
Enzyme induction secondary effect of insecticides, 139
Epitremerus pyri Nalepa, 57
EPTAC *see* Eptam
Eptam (EPTC) crop uses, 4
Erysiphales sp., 32,33
Ethazol soil use, 30
Ethirimol *also see* pyrimidin fungicides:
mammalian toxicity, 33,38
Ethylenethiourea:
carcinogen derived from dithiocarbamates, 29,30
Excessive toxic insecticide residues:
regarding need for insect selectivity, 53

F

Fatty acids, in cuticle tanning, 91
Fatty acid amines, as insect cuticle dessicants, 92
Feeding deterrency, insect, 78
Fenazaflor:
acaricidal activity, 156,164, 165
mechanism of toxic action, 171
structure of, 160,170
toxicological selectivity, 165
Fenitrothion, as replacement for parathion, 104,105
Fenson:
acaricidal activity, 156
structure of,160
Fentin acetate, as insect antifeedant, 83,84

Fentin:
fungicide action, 24,26
mammalian toxicity, 29
Fenthion (Baytex), 122,126
Ferbam:
fungicide action, 24
mammalian toxicity, 29
Fluenethyl:
acaricidal activity, 156,164, 165
mechanism of toxic action, 169,170,171
structure of, 160,170
structure-activity relations, 170
toxicological selectivity, 165
Folpet:
as mutagen, teratogen, 30
fish toxicity, 29
Formaldehyde, 30
Formamidine acaricides *see* C-8520, H-20013, chlordimeform
Formetanate:
acaricide activity, 156,164, 165
structure of, 158
toxicological selectivity, 165
Formparanate:
acaricidal activity, 156,164, 165
structure of, 158
toxicological selectivity, 165
Fumigants:
conditions for use, 30
general toxicity, 30
non-selectivity, 31
Fungicides:
basis for selective toxicants, 24
classed as protectants or systemics, 22, 23
developmental costs, 42
inorganic compounds as, 23
legal requirements affecting development, 42
mammalian toxicity, 29
need for adequate screening procedures, 42
non-target selectivity, 28, 29,30
pathogen and host plant selectivity, 24,25,26,27,28
pathogen resistance, classification by mechanism, 35
phytotoxicity of, 25,26
rate of fungal uptake, 25,26
selectivity:
chitin inhibition vs. plant and animal toxicity, 37
example of a site specific mechanism, 37
soil fungicides:
comparison to surface protectants, 30
protectants and disinfectants, 30,31
volatile disinfectants or fumigants, 30
surface protectant fungicides:
alkylation of cellular sulfhydryl group by, 27
as general toxicants, 26
chelation of metal-sensitive enzymes by, 26
classification of, 23,24
fish toxicity of, 29
enzyme inhibition in energy production by, 27,28
principles of use, 23
resistance to, 28
systemic fungicides:
accum. and plant degradation products, 30
advantages of, 43
anti-bacterial agents, 32, 33
basis for selective toxicants, 34,35,36,37
cellular sites and mech. action,36
characteristics of usage, 39,40
chemotherapeutants, 31,32, 33,34,35,36

classification of, 37,39
development of pathogen resistance, 34,35
differential tox. plant/pathogen, 31
differential uptake, 35
foliar sprays, 40
mammalian hazards, 37,38
non-target toxicities, 36, 37,38,39
pathogen and host selectivity, 32,33,34,35,36
phytotoxicity of, 33
protection of plant propagules, 39
resistence development, policies to minimize, 43
resistant pathogens as major limitation to use, 43
soil treatment, 39,40
translocation within plant, 32

G

Genite:
acaricidal activity, 156
structure of, 160
Glyodinryania in apple pest management, 56
Griseofulvin effect on insect cuticle, 88,89

H

H-20013 (Formamidine acaricide), 161
Hansch analysis, 95
Heliothis virescens Fab., 60,61
Heliothis zea Boddie, 60,61
Helminthosporium maydis (corn leaf blight), 22
Heptachlor, 69
Herbicides:
broad spectrum, 15
development costs, 5,6
hazard to animals, 15
in environment management, 16
lack of environmental justification for curtailment, 16,17
persistence, need for, 3,9
registration problems, 20
safety, 2,9
screeening tests, 6
selective, definition of, 3
Hexamethylditin *see* Pennwalt TD-3052:
insect antifeedant, 81,82
structure of, 80
Hinosan, 36,37
1-Hydroxyethylene glyoxalidines fungicidal activity, 25
Hydra littoralis, 144

I

Insect antifeedants:
general advantages of, 82
mechanism of action, 83,84
systemic action of, 82
Insect control:
biological, 12
contrast to weed control, 15
demand for complete mortality in, 50
integrated pest control, 12, 18,52
reliance on chemicals, 50
Insect hormones, as possible targets for metabolic toxicants, 86,87
Insecticide application:
as factor in selective control, 48,50,51,58,59
efficiency and precision in, 60
wind drift in, 60
Insecticides:
accidental poisoning, man, 71
antifeeding compounds, 78-85
classification of toxic hazards of, 68,69,70,71

effects on non-target organisms, 135-146,150
environment, effects on, 12
environmental persistence of, 71,73
insect resistence and, 12
mainstay of pest control, 49
mammalian toxicities of, 69
pest management index vs. mammalian toxicity, 68,69, 70
practical methods of improving selectivity in usage, 59,60,61,62
predatory and beneficial insects, effects on, 12
physiological basis for, 49, 74
trends in use of hazardous insecticides, 71,72

Insecticides, selective:
advantages in fruit pest control, 56,57
behavioral toxicants as, 77, 85
chemical sterilants as, 61
cuticle waterproofing and, 92,93
definition of, 47,48,75
ecological selectivity of, 59, 74
insect pathogens with, 62
metabolic toxicants and, 85-95,136:
α-glycerophosphate shuttle, as target, 86
hemolymph biochemistry, as target, 86
insect hormones, as targets, 86,87
insect sterols, as target, 86
integument, as target, 87, 88,89,90,91
monoselectivity (single species) of, 77
physiological basis for, 49, 74
realistic usage goals and economic justification of, 52,53,54,55,73,74,75,104, 105
required for integrated control, 49
resistance development and need for, 53
secondary and tertiary toxic effects of, 140
strategies in design of, 75, 76,95-104
specific biotic insecticides, as, 62
unfavorable economics of, 68
usage conditions pertinent to obtaining selectivity, 48,50,51,58,59
via capitalizing on behavioral characteristics, 60, 61
via differential toxicity, 56
via timed application, 56
via traps with attendant, 61

Insects:
behavior, 78
beneficial species, 15
pheromones, 62
predator species, 15

Invertebrates (non-arthropod), comparative biochemistry, need for, 144,145

J

Juglone (5-hydroxy-1,4-napthoquinone):
insect antifeedant, 79,80,83
mechanic of action, 83
structure-activity relationship of napthoquinone analogues, 83

Juvenile hormone as target for metabolic toxicants, 86

Juvenile hormone analogues and cuticle tanning: 91
selectivity, 86
economics of possible usage, 86

K

Karmex (diuron) crop uses, 4
Kasugamycin, 32,33,36
Kitazin:
  mechanism of action, 36
  mammalian toxicity, 38
  selectivity of, 33

L

Laspeyresia pomonella (codling moth), 56
Lasso (alachlor) crop uses, 4
Locust, 102
Lumbricus terrestris, 144
Lymnea palustis, 144

M

"Magic meta methyl," 96
Malathion, 69
Maneb:
  as cuticular insecticide, 89, 91
  fungicidal activity of, 24, 27,29
MBC (Methylbenzimidazolecarbamate) *see* benzimidazoles and thiophonates):
  basis for selective toxicity, 34
  mechanism of fungicidal action, 34
Meliaceae sp., 80
Meliantrol, insect antifeedant, 80
Mercenaria mercenaria, 144
Mercurial fungicides, mammalian toxicity, 29
Methyl bromide, 30
1-methyl-2-mercaptoimidazole, as cuticular tanning inhibitor, 90
methyl parathion:
  cost vs. selective action, 55
  insecticidal activity, 69, 96,97,98,99,100,126
  vs. ethyl parathion, 55
6-methyl-thiouracil, insecticidal action, 90
6-methoxybenzoxazolinone, insect antifeedant, 79,80
Methoxychlor, 69
Mevinphos:
  acaricidal activity, 156, 164,165
  structure of, 157
  toxicological selectivity, 165
Mexican bean bettle, 103
Microsomal oxidases:
  of non-arthropod invertebrates, 144
  histological detection of, 145
Milbex:
  acaricidal activity, 156
  structure of, 160
m-isopropylphenyl-N-methyl carbamate *see* alkylphenyl carbamates
Mites:
  predator-prey ratio, 58
  predaceous, control by selective insecticide, 58
  sensitivity to insecticide spray program, 58
  terminology and classification, 155,156
Mitosis, inhibition in fungi, 36
Model ecosystem, value in assessing pesticide biodegradibility, 142
MON-0585, insecticidal activity, 89
Morpholines, fungicidal activity, 32,33
Mosquito larvicides and insecticides:
  larvicide testing of, 119, 120
  market estimates for, 128
m-sec. butylphenyl-N-methylcarbamate *see* alkylphenyl carbamates

Musca domesticus (housefly), 91,98,99,102,104,170
Mussel (*Anodonta* sp.), toxicant metabolism by, 144
Mutagenicity, as tertiary insecticide action, 137,138
Mrak committee, recommendations of, 1,2,3,8,11
Mylone, 30

N

Nabam, fungicidal activity, 27
Naled:
  acaricidal activity, 156
  structure of, 157
Neem tree principle *see* azadirachtin
Nemagon, fungicidal activity, 30
Neurospora crassa, 25, 37
Nissol:
  acaricidal activity, 156, 164, 165
  mechanism of toxic action, 170,171
  structure of, 160,170
  toxicological selectivity, 165
N-methyl carbamates, structure-activity factors in comp. toxicity, 100,101,102,103, 104
Non-target organisms:
  imperfect selective tox. of insecticides and, 136-141
  secondary effects of insecticides and, 137
  tertiary effects of insecticides and, 137
  variation of toxicant metabolism in, 142
Nutsedge (*Cyperus rotundus* L.):
  weed problem, 5
  herbicide controls, 5
Nystatin, fungicidal activity, 36

O

Omite:
  acaricidal activity, 156
  structure of, 160
Organophosphate insecticides:
  as phosphorylating and alkylating agents, 138
  p-acetylphenyl phosphate esters, methylation of and selectivity, 97,98.99,100
  reactivity of degradation products, 143
Orius spp., 61
Ovex:
  acaricidal activity, 156
  structure of, 160
Oxathiins *see* carboxin
Oxine (8-hydroxy-quinoline), fungicidal activity, 26
Oxythioquinox:
  acaricidal activity, 156, 164,166,167,168
  mechanism of toxiaction, 172
  structure of, 162,170
  toxicological selectivity, 166,167

P

p-acetylphenyl phosphate esters, effect of methylation on insect selectivity, 97,98,99,100
Parathion:
  acaricidal activity, 156, 164,165
  biomagnification in non-target organisms, 142
  cost vs. methyl parathion selective toxicity, 55
  insecticidal activity, 69, 71,123,124,125,126
  metabolism by earthworm, 144
  microsomal oxidation of, 145
  structure of, 157
  toxicological selectivity, 165

Patent procedures, 2
PCNB:
  mammalian toxicity, 29
  soil use of, 30
Pectinophora gossypiella (pink bollworm), 54,55
Pentac:
  acaricidal activity, 156
  structure of, 158
Pennwalt TD-3052 *see* Hexamethylditin
Periplaneta americana, 121
Pest control, strategies in, 12,13,14,18,50,51
Pesticide chemical industry:
  government subsidies or assistance, 2
  incentives for development of selective chemicals, 1, 2,8
Pesticides:
  development cost, 1,6,7, 129
  govt. program to reduce level of use, 17
  hazards to man and environment, 2
  public concern over, 53
  relationship to herbicides, 17
  selective action through use, 48,49
  selectivity based on differential physical selectivity, 48,49
Pests, class distinctions among, 13,14
1-phenyl-2-thiourea, as cuticular insecticide, 89,90
Phorate:
  acaricidal activity, 156
  structure of, 157
Phycomycetes, systemic fungicide selectivity, 32,33
Physa sp., 144
Phytophthora infestans, potato late blight fungus, 21
Plant breeding for insect resistance, 78
Plant chemotherapy, definition, 31
Plant disease:
  chemical control, history of, 22
  chemical protectant and/or eradicant of, 22
  crop losses to, 21
  resistant varieties and, 22
Plictran:
  acaricidal activity, 156, 160,164,166,167,168
  insect antifeedant, 82
  structure of, 80
  toxicological selectivity, 167
Polyoxin:
  effect on chitin of fungi and insects, 88,89
  mechanism of action, 36
  mammalian toxicity, 38
Polyphenoloxidase (in insect cuticle), stimulation by benzoylphenylureas, 93
Propargite *see* Omite
Protein synthesis, inhibition in fungal ribosomes, 36
Protoplast membrane, permeability altered by nystatin, 36
Psylla pyricola Foerster, 57
Pyrethroids, synthetic, 96
Pyrimidine fungicides, 32,33
Pyrazophos:
  fungicidal activity, 22,27, 28
  mammalian toxicity, 38

Q

Quinones as insecticide target in insect integument, 87

R

R-17335:
  acaricidal activity, 156
  structure of, 158

R-10044, acaricide structure, 160
RE-11775 (m-sec. butylphenyl-N-benzenesulfenyl-N-methyl carbamte) *see* alkylphenyl carbamtes:
  economic factors in development, 128,129,130
  mosquito larvicide, 123,124, 125,126
  failure of public development, 129,130
  insecticidal activity, 120, 121,122
  mammalian toxicity, 121,122
  manufacturing considerations, 127
  metabolism of, 127
  synthesis of, 117,118
  toxicity to resistant mosquitos, 123,124,125
Reproduction, birds, effects of chlorinated insecticides, 139
Rotenoids, selective action of, 140

S

Saccharomyces pastorianus, 25,37
Sarcophaga bullata, 91
SD-17250:
  acaricidal activity, 156
  structure of, 158
Secondary pest outbreaks, need for insecticide selectivity, 53,54
Selective Inhibitory Ratio (SI-R), insecticide rating by, 97
Selective toxicity:
  metabolic toxic effect in, 76,77
  receptor affinity and, 76,96, 97
  via penetration and distribution in the body, 76
Selective Toxicity Ratio (STR), insecticide rating by, 97
Silica aerogels, as insect cuticle dessicants, 92
Snail (*Lymnaea palustis*), toxicant metabolism by, 144
Sparrow, 102
Spider mites:
  acaricide selectivity, 170, 172
  outbreak after carbaryl use on cotton, 54
Spodoptera littoralis, 84
Sterol synthesis, inhibition in fungal endoplasmic reticulum, 36
Streptomycin, mammalian toxicity of, 32,33,36,38
Structure-activity analysis, 95
Synergistic ratio, use in assessing oxidative degradation of insecticides, 141,142

T

Target specific pesticides:
  definition and qualifications of, 113,114
  possibilities favoring development of, 130,131
TEPP:
  acaricidal activity, 156
  structure of, 157
Teratogenicity, as tertiary insecticide action, 137
Tetradifon:
  acaricidal activity, 156
  structure of, 150
Tetranychus mcdanieli McGregor, 57
tetranychus urticae Koch, 57. 58,167
Tetrasul:
  acaricidal activity, 156
  structure of, 160
Thiobendazole:
  benzimidazole fungicide, 33

mammalian toxicity, 38
mechanism fungicidal action, 36
Thiolcarbamate herbicides, in nut sedge control, 5
Thiophonate fungicides, 32,33
Thiouracils, cuticular insecticides via phenoloxidase inhibition, 90
Thiourea acaricides *see* C-9140
Thioureas, cuticular insecticides via phenoloxidase inhibition, 90
Thiram, 26
Ticks, terminology of, 155
Toxaphene, 69
Treflan, trifluralin, crop uses of, 4
Triamiphos, mammalian toxicity of, 37
Triarimol, undesirable toxicity of, 38
2,4,6-trichlorophenoxyethanol:
as insect antifeedant, 81,83
structure of, 80
Tridemorph:
mammalian toxicity of, 38
morpholine fungicide, 33
Trifluralin *see* Treflan
Triforine:
piperazine fungicide, 33
mammalian toxicity of, 38
mechanism of action of, 36
Triphenyl tin *see* Fentin
Typhlodromus occidentalis Nesbitt, 56,58

U

U-36059:
acaricidal activity, 156,164, 166,167,168
toxicological selectivity, 167
UC-41305:
acaricidal activity, 156,164, 166,167,168
structure-activity aspects of, 171
structure of, 162,170
toxicological selectivity of, 166

V

Vapam, 30
Viral encephalitis, mosquito vectors of, 123
Vorlex, 30

W

Weed control:
biological control in, 18
cultivation in, 3,11
contrast to insect control, 15
expansion of checked species after control method change, 17
integrated control in, 18
national strategy for, 19
selective control in, 11
Weeds:
classification of, 4
herbicide resistance, development of, 16,17
mobility of, 14
types per crop field, 5
Weed science:
role in national pest control strategy, 14
weed control technology, 14
Weed Science Society, weed classification by, 4
Weed scientists, role in national pesticide policies, 18

Z

Zectran, toxicity modification by acetylation, 101
Zineb:
as cuticular insecticide, 91
fungicidal activity of, 24,27
Ziram, as cuticular insecticide, 90